IDD

IN

LIVESTOCK

IDD
IN
LIVESTOCK

Ecology & Economics

Edited by

DR CHANDRAKANT S. PANDAV

and

DR A.R. RAO

DELHI
OXFORD UNIVERSITY PRESS
MUMBAI CALCUTTA CHENNAI
1997

Oxford University Press, Walton Street, Oxford OX2 6DP

Oxford New York
Athens Auckland Bangkok Calcutta
Cape Town Chennai Dar es Salaam Delhi
Florence Hong Kong Istanbul Karachi
Kuala Lumpur Madrid Melbourne Mexico City
Mumbai Nairobi Paris Singapore
Taipei Tokyo Toronto

and associates in

Berlin Ibadan

ISBN 0 19 564205 8

This book is published in collaboration with the International Council for Control of Iodine Deficiency Disorders

Cover design : Gauri Shankar Mukherjee

Printed at Gopsons Papers Ltd., Noida 201301
and published by Manzar Khan, Oxford University Press
YMCA Library Building, Jai Singh Road, New Delhi 110001

**TO THE GLOBAL PARTNERSHIP
DEDICATED TO THE ELIMINATION OF
IODINE DEFICIENCY DISORDERS**

An Ancient Scourge of Mankind

The People of the affected countries
The Governments of the affected countries
The Salt Producers of each country

The International Agencies-especially
The World Health Organization
The United Nations Children's Fund
The World Bank
The Micronutrient Initiative
Kiwanis International
Program Against Micronutrient Malnutrition

The International Expert Network of
International Council for Control of
Iodine Deficiency Disorders
(ICCIDD)

The Bilateral Agencies especially
The Australian Agency for International Development
The Canadian International Development Agency
The Netherlands Ministry for Development Cooperation
The Swedish International Development Agency
The United States Agency for International Development

The International Council for Control of Iodine Deficiency Disorders (ICCIDD) is a non-profit non governmental organization dedicated to the sustainable elimination of Iodine Deficiency Disorders (IDD) throughout the world. The ICCIDD was granted an official status as an International NGO at the 47th World Health Assembly held in Geneva in 1994. It's activities are supported by donations/ grants from the Australian Agency for International Development (AusAID), the Canadian International Development Agency (CIDA), the Micronutrient Initiative (MI), the Netherlands Ministry for Development Cooperation, the Swedish International Development Agency (SIDA), the United Nations Children's Fund (UNICEF), the United States Agency for International Development (USAID), the World Bank, the World Health Organization (WHO), and others.

Contents

Foreword

Iodine deficiency as a cause of losses in livestock production was well known to an earlier generation of scientists and agriculturalists. Reduced fertility, abortion and stillbirth, with defective growth are the avoidable consequences of subnormal thyroid hormone levels in cattle, sheep, pigs and chickens. Defective milk production and fish production are the other well known features of iodine deficiency.

Increases in livestock productivity will definitely occur with the correction of iodine deficiency thus providing an increase in the economic return.

The editors of this book on 'Iodine Deficiency in Livestock' have presented a timely multidisciplinary report on the proceedings of a Technical Review Meeting held in Uttar Pradesh, India on 27-8 January 1995. The book has been published by Oxford University Press in Collaboration with the SE Asia Regional Office of the International Council for Control of Iodine Deficiency Disorders (ICCIDD).

The ICCIDD is an expert multidisciplinary network of 400 professionals from more than 80 countries which is officially recognized by the World Health Organization as a non-government organization (NGO).

Iodine Deficiency Disorders (IDD) are one of the most common causes of preventable brain damage in the world today. Recognition of the fact has led to a global programme which has been led by WHO, UNICEF and the ICCIDD in cooperation with governments of more than 80 developing countries.

The Indian government alongwith the governments of many other countries has adopted the policy of Universal Salt Iodization (USI) which requires that all salt for human and animal consumption be iodized.

This policy has been adopted as Iodine Deficiency Disorders are a significant barrier to the development in India and many other countries. This applies to both human and animal development and, therefore, is of great importance.

This book reviews the status of the USI programme in India and the current status of the prevention and control of IDD. It provides a very useful reference for the progress of a remarkable programme which has the goal of eliminating IDD in both human and animal populations by the year 2000.

Massive benefits are to be gained from the achievement of this goal. Moreover remarkable progress has already been achieved.

We commend this very useful book to a wide audience to increase awareness of this great opportunity for improvement in animal productivity and human well being.

Basil Hetzel
Chairman
ICCIDD

F. Delange
Executive Director
ICCIDD

Preface

Iodine Deficiency Disorders (IDD) in animals is yet to receive the same attention as that of human beings. Possibly, this is a fall out of the view that health of animals does not affect us directly. This view has changed over the years, and we now, especially in India, realize the important role the health of livestock plays on our health. Over 20 per cent of the income from the agricultural sector is created by livestock, over 70 per cent of rural households depend on animals either for a livelihood or for supplementing an income. If by reducing the incidence of IDD in animals, we can improve their health and increase productivity, we may in turn tip the scales in favour of better health and well being for human beings.

The extent of our dependence, often existence itself, on animals was underlined by an incident that took place years ago.

As a young medical intern from the All India Institute of Medical Sciences, New Delhi posted at the Comprehensive Rural Health Services Project, Ballabgarh (Haryana, India) an incident occurred that remains sharply etched in my memory - that of Hardevi, a mother whose three children were suffering from pneumonia. I was the only doctor on duty. I admitted all her three children to the hospital and requested her to stay back to care for them. In general, the hospital was always crowded and the sanctioned hospital staff could never cope with the ever increasing load of patient care. Therefore, we had to seek the help of the patient's relatives and their assistance in taking care of the basic nursing needs. Hardevi said that she could not stay in the hospital as she had to return home 5 kilometers away, to take care of her buffalo,who incidentally was also sick. She could not risk her buffalo dying, for the lives of her three children, as well as the survival of her husband and his parents depended on the buffalo. We pleaded with her that her children would die if she took them away, but Hardevi stood firm. She took them back to her house.

One by one the three little children died. The buffalo survived. It was a harsh lesson in the health economics of survival, and the well being of a buffalo overriding the lives of one year old Satpal, three year old Hira and five year old Dayal. I still remember her words, " Doctor Saab, I am a mother. I will of course feel very sad if my children do not survive. But if my buffalo dies, the whole family will starve to death. I can always have children later on!".

During a visit to China in June 1993, I chanced upon a picture of a Chinese family in a national daily. The family was shifting to a poultry farm at the height of summer with temperatures soaring between 40 - 45°C. They had a young, one year old child who was clearly suffering from the discomforts of shifting residence in such weather, but the parents were more concerned about the welfare of the hens and chickens they were taking along with them. The only fan they possessed had been placed in the poultry farm. These birds were the source of their livelihood. Without them, the survival of the entire family and not just of one child, was at stake. Poultry gained primary economic importance over the young child.

That the health of animals, is directly related to the health of human beings, was driven home forcefully, by the strategy adopted to achieve a 100 per cent coverage rate in the IDD elimination programme launched in Ecuador. It had achieved an 80 per cent coverage rate mainly through the effective use of radio broadcasts on the importance of consuming iodized salt. But by the 1990s, coverage had tapered off, eventually reaching a plateau. No extra persuasion worked, until health scientists instituted an awareness programme on the benefits of iodizing salt for animal consumption. Soon the programme achieved its goal and is maintaining it — by virtue of improving the health of Ecuador's animals!

Many of us involved in IDD prevention, control and elimination programmes felt the need for a holistic approach to a medical problem that had geographical, social and economic dimensions. The Technical Review Meeting of Iodine Deficiency Disorders in Livestock Populations, held at G.B. Pant University for Agriculture and Technology, Pantnagar, Uttar Pradesh, India was the culmination of efforts taken over the past two decades to focus attention on IDD in animals and the role that iodine sufficient animals play in the well being of human beings.

'IDD in Livestock – Ecology and Economics', is a forerunner in the multi–disciplinary approach to viewing and creating intervention methodologies in achieving the goal of eliminating IDD from the world and sustaining its elimination. It contains a selection of papers presented by medical and veterinary scientists, pediatricians, nutrition experts, agriculture and soil scientists and entomologists.

I thank the contributors Drs B.R. Shamanna, K. Anand, N.K. Arora, M.G. Karmarkar, S.S. Khirwar, K. Pradhan, V. Kapoor, A.R. Rao, S.P. Pachauri, James A. Pittman (Jr), Constance S. Pittman, S.P.S. Bedi, Rita Singh Raghuvanshi, S.S. Sachan, T.R. Rathore and M.G. Venkatesh Mannar. This book would not have been possible without their expertise.

I appreciate the support given by the International Council for Control of Iodine Deficiency Disorders (ICCIDD). I am grateful to Dr Basil S. Hetzel, Chairman and Dr Francois Delange, Executive Director, ICCIDD for their foreword and comments on the text.

The continuing work in the field of Iodine Deficiency Disorders has been possible with unstinted support and keen interest of Prof. Lalit M. Nath, former Director, All India Institute of

Medical Sciences, Professor and Head, Centre for Community Medicine, and all the faculty and staff members of the Centre.

I was trained as an International Clinical Epidemiology Network (INCLEN) Fellow (Health Economics; Level - II) at the Department of Clinical Epidemiology and Biostatistics and Centre for Health Economics and Policy Analysis (CHEPA), at McMaster University, Hamilton, Ontario, Canada. I would like to place on record my acknowledgement to the INCLEN Programme and to the London School of Hygiene & Tropical Medicine where I was trained in 'Measurement & Assessment in Health & Nutrition Interventions in Developing Counbtries'.

I would like to acknowledge the UNICEF offfices of SAARC member countries, UNICEF Regional Office, Kathmandu and H.Q. office New York and WHO Regional Office, New Delhi and H.Q. Office, Geneva for their continued co-operation and partnership for sustaining elimination of iodine deficiency disorders.

This book would not have been possible without the support and keen interest of colleagues at the All India Institute of Medical Sciences and ICCIDD, New Delhi, Dr M. G. Karmarkar, Dr N. K. Arora, Dr Shamanna, Dr K.Anand and Ms Smita Pandav. Dr Pralhad Rao from National Institute of Nutrition, Hyderabad initiated and took the lead in sharing the responsibility of the scientific programme of the meeting.

Mr Rashid Ahmed, ICCIDD Project Officer worked laboriously in giving context and shape to the manuscript. My special thanks to him.

I am thankful to Ms Veena and Mr Ashok Negi for office support and to Mr Pritam Singh Tanwar for computer assistance.

I am also indebted to Dr Rajesh Pandav and Ms Vandna Khare for their comments and suggestions while compiling this book and to Mr Gauri Shankar Mukherjee for the cover design.

It has been a pleasure to work with Dr A.R. Rao, former Professor of Physiology, College of Veterinary Sciences, Haryana Agriculture University, Hisar as a co-editor.

I also wish to thank the Oxford University Press for assistance with the final publication.

Dr Chandrakant S. Pandav

Regional Co-ordinator,
ICCIDD
(South Asia & Pacific Region)
New Delhi, India

Additional Professor,
Centre for Community
Medicine, AIIMS
New Delhi, India

List of Authors

Dr Chandrakant S. Pandav
Regional Coordinator,
South-Asia & Pacific
International Council for
Control of Iodine
Deficiency Disorders,
Additional Professor,
Centre for Community
Medicine (CCM),
All India Institute of Medical
Sciences (AIIMS),
New Delhi, India

Dr B.R. Shamanna
Senior Resident,
CCM, AIIMS,
New Delhi, India

Dr K. Anand
Assistant Professor,
CCM, AIIMS,
New Delhi, India

Dr N.K. Arora
Associate Professor,
Department of Pediatrics,
AIIMS,
New Delhi, India

Dr M.G. Karmarkar
Senior Advisor,
International Council for
Control of Iodine
Deficiency Disorders
New Delhi, India

Dr R. Prakash
Salt Commissioner to the
Government of India,
Salt Department,
Ministry of Industry,
2A, Lavan Marg,
Jhalana Dungari,
Jaipur, India

Dr N. Pralhad Rao
Sr Deputy Director
National Institute of
Nutrition,
Indian Council of Medical
Research,
Hyderabad, India

Dr G.N.V. Brahmam
Sr Research Officer
National Institute of
Nutrition,
Indian Council of Medical
Research,
Hyderabad, India

Dr S.S. Khirwar
Associate Director Research
(Veterinary and Animal
Science),
Directorate of Research,
Chaudhary Charan Singh
Haryana Agricultural
University, Hisar, India

Dr K. Pradhan
Vice-Chancellor,
Orissa University of Agriculture and Technology,
Bhuvaneshwar, India

Dr V. Kapoor
Department of Animal Nutrition, CCS,
Haryana Agricultural University, Hisar, India

Dr A.R. Rao
Department of Veterinary Physiology, CCS
Haryana Agricultural University, Hisar - 125004

Dr S.P. Pachauri
Professor and Head of Department of Medicine,
College of Veterniary Sciences,
G.B. Pant University,
Pantnagar, India

Dr James A. Pittman
University of Alabama
School of Medicine
Birmingham, Alabama,
U.S.A

Dr Constance S. Pittman
University of Alabama
School of Medicine
Birmingham, Alabama,
U.S.A

Dr S.P.S. Bedi
Animal Nutrition Division,
Indian Veterinary Research Institute,
Izatnagar, India

Dr Rita Singh Raghuvanshi
Department of Foods and Nutrition,
College of Home Science,
G.B. Pant University of Agriculture & Technology,
Pantnagar, India

Dr T. Prasad
Division of Dairy Cattle Nutrition,
Dairy Research Institute,
Karnal, Haryana, India

Dr K.S.N. Prasad
Division of Dairy Cattle Nutrition,
National Dairy Research Institute, Karnal, Haryana, India

Dr R.S. Sachan
Department of Soil Science,
G.B. Pant University of Agriculture and Technology,
Pantnagar, Nainital, U.P, India

Dr T.R. Rathore
Department of Soil Science,
G.B. Pant University of Agriculture and Technology, Pantnagar,
Nainital, U.P, India

Dr M.G.V. Mannar
Executive Director,
The Micronutrient Initiative,
250 Rue Albert Street
Ottawa, Canada

Lt Col R. Sankar
Classified Medical Specialist
Department of Thyroidology
INMAS, Lucknow Road
Timarpur, Delhi, India

Introduction

Although manifestations of Iodine Deficiency Disorders (IDD) have been recognized since ancient times in human beings, IDD in animals has only recently received attention. Most of our knowledge of the effects of iodine deficiency comes from medical research using experimental animals in order to understand the role of iodine in development, homeostasis and reproduction. Livestock health and productivity, however, are also affected by environmental iodine deficiency, much the same way as with humans.

It is generally believed that those geographical locations where IDD is more pronounced in human beings are likely to be the same as for animals (such as sheep, goats, cows, buffalo, pigs, and chickens) due to the ecology of iodine. A lack of iodine in the diet of farm animals has been shown to substantially reduce the yield of eggs, milk, meat, agricultural draught power, and dung for cooking fuel and maintaining soil fertility. This is especially relevant to rural people given the intimate interrelationship in the health of both humans and animals (Fig. 1).

Given the extent to which iodine deficiency disorders affect livestock population, and consequently human health and the economy, a review meeting was deemed necessary to consolidate a body of knowledge on the subject. A multidisciplinary group consisting of economists, soil scientists, nutritionists, medical doctors, policy makers, biochemists, and veterinarians met

TUESDAY FEBRUARY 7 1995, THE TIMES OF INDIA, NEW DELHI 17

HEALTHFILE

People keep as well as the animals they live with

By Kalpana Jain

Thirty-five-year-old Hardevi went to a primary health care centre with her three children suffering from pneumonia. The doctor on duty asked her to immediately admit the children to the centre. Hardevi was asked to stay back and help take care of the children as staff was scarce.

Hardevi, despite the serious condition of her children, said she could not do so. Back in her village, some five km from this centre, her buffalo was also lying ill. And she couldn't risk the death of the buffalo for the lives of her other three children, her husband and his parents depended on the survival of the animal.

The doctor pleaded with the woman. He explained that if she took them away they would all die. Hardevi did not relent. She took the children back. One by one her three children -- two-year-old Satpal, four year-old Hira and six-year-old Dayal died. The buffalo survived.

This incident occurred sometime back at the primary health care centre at Ballabgarh run by the All India Institute of Medical Sciences. But more shocking is the fact that this is not an isolated case. "Such incidents are common, " says associate professor of pediatrics N.K. Arora explaining that the health of the entire food chain is vital. Health of soil, plants and animals is what contributes to the overall health and growth of a child, he says.

The economic consequences of diseases of livestock are important while considering public health measures, especially for the rural areas where dependence for livelihood is entirely on livestock. The yield from this livestock affects others who are not directly dependant on livestock for economic sustenance. It affects health depending on the nutrients given to the livestock through plants and other fodder. To achieve this complete food chain the health of soil and plants has also to be ensured.

"The economic consequences of livestock ailments are acute. And it is important to look after the health of livestock if the health of human beings is to be taken care of," says Dr C.S. Pandav, who for the first time brought medical scientists handling issues of health of human beings and veterinary scientists together.

The two-day meeting was held at Pantnagar's Agricultural University. Participants included experts from the National Institute of Nutrition,paediatricians, medical scientists from the All India Institute of Medical Sciences, agriculture scientists, soil scientists,and entomologists.

These scientists are approaching health in its totality which means healthy soil, plants and animals. The soil where plants grow should have adequate amounts of all nutrients.

Animals, who are also feeding on the same plants that are deficient in micronutrients suffer in several ways. The effect of iodine deficiency in farm animals has been studied extensively. The rate of miscarriage is high, says Dr Pandav. No abortions occurred in goats when their intake of iodine was increased ten times. The rate of miscarriage in these goats was as high as 47 per cent.

It was found that iodine deficiency leads to several other disorders. The calves may be born blind, hariless, or extremely weak. The rate of conception goes down and there is infertility as well. The yield of milk and wool is also reduced substantially.

In those parts where iodine deficiency disorders are common amongst human beings, they are known to affect animals as well. But in places where iodised salt is available for human beings, uniodised salt is kept for giving to livestock, adds Dr Pandava, the regional coordinator for iodine disorder control for south- east Asia.

Experts feel that with this step scientists are moving ahead of superspecialisation. They are now connecting disciplines and completing a chain to arrive at better solutions to health care.

Fig. 1. Newspaper article illustrating the 'real' effects of IDD in livestock.

on 27 and 28 January, 1995 at the G.B. Pant University for Agriculture and Technology in Pantnagar, Nainital district, Uttar Pradesh, in an effort to diversify efforts to eliminate iodine deficiency disorders throughout the country. This publication represents the proceedings of 'The Technical Review Meeting on Iodine Deficiency Disorders in Livestock'.
The book is divided into seven parts:

Part I Iodine Deficiency Disorders — An Overview

Part II Livestock and the Indian Economy

Part III Iodine Deficiency Disorders in Livestock: Overview

Part IV Feeds and Fodders — Iodine Status

Part V Iodine Deficiency Disorders in India: Animal Studies

Part VI Control Measures for Iodine Deficiency Disorders in Livestock

Part VII Recommendations of the Technical Review Meeting on Iodine Deficiency Disorders in Livestock

The articles have been arranged serially to provide a logical argument for focusing attention on iodine deficiency in livestock populations; however, each article can easily stand independently. Fig. 2 serves as a conceptual model for the reader to consider while reading the contents of this report. It emphasizes a holistic approach in examining how IDD affects both humans and animals through the food chain, starting at the environmental source and moving upwards in an ecological fashion. A series of specific recommendations made at the meeting concludes the publication.

The Ecology of Iodine - a Review

The iodine cycle in nature affects the distribution of this element in soil, plants, and animals within a system. Several articles review the mechanism by which iodine deficiency results from bio-geographical, rather than social and economic conditions. IDD in animals may also be more severe due to a relatively local food chain, whereas human beings have access to a variety of foods imported from non-deficient areas.

Effect on People	**Humans** ⇧	*Health and socio-economic impacts*
Effect on Animals	**Livestock** ⇧	*Clinical and reproductive disorders, decreased productivity*
Low availability of iodine	**Plants**	*Iodine-poor feeds and fodders, goitrogens*
	⇧	
	Water, Soil	*Environmental iodine deficiency*

Fig. 2. A hierarchical food chain where iodine deficiency is attenuated at each trophic level with consequences.

Availability of Iodine

An understanding of the causes of IDD in livestock first requires an examination of the 'environment' which precipitates the disease. This approach emphasizes an understanding of the root etiology of IDD in livestock and its prevention, as opposed to a purely medical approach.

The amount of iodine available to animals from vegetation is generally affected by three factors:

1) iodine content of water, soil and plants;
2) presence of goitrogens (antimetabolites); and
3) parasitic infection

1) Iodine Content of Water, Soil and Plants

Unlike other nutrients, like iron, calcium or vitamins, iodine does not occur naturally in specific foods; rather, it is present in the soil and is imbibed through foods grown on that soil. As a result, animal and human populations which are totally dependent on food grown in such soils are unable to obtain the physiological amounts of iodine that they need and deficiency symptoms appear.

The abundance of iodine and other trace elements in soil and plant systems is a complex problem influenced by various factors: soil type and pH, local geology, type of fertilizers used, source of irrigation, average annual precipitation, crop species and rotation pattern, stage of plant maturity, and climatic factors. The chapters on 'Effects of iodine deficiency on livestock',

'Iodine profile of foods and foliage', and 'Iodine status of feeds and fodders', also consider these aspects.

2) Goitrogens

Goitrogens are chemical substances found in certain plants that interfere with normal iodine metabolism and can precipitate IDD despite an otherwise adequate iodine intake. They are a secondary factor to dietary iodine deficiency, which is more common. Goitrogens are easily counteracted with sufficient dietary iodine supplementation. Three general categories of goitrogens exist which affect of animals: i) cyanogenic glucosides, ii) mimosine, and iii) other iodine antagonists (i.e., several inorganic elements). Many articles in this book discuss goitrogens; 'Feed antimetabolites which influence thyroid function in livestock' is one.

3) Parasitic infection

Hypothyroidism in animals may also be aggravated by parasitic infection. Histopathological evidence has demonstrated sclerotic lesions in the thyroid glands of cows due to Onchocerca infections.

The effect of intestinal worms in decreasing iron absorption in humans is perhaps analogous to iodine uptake in parasitized animals.

Effects on Animals

Iodine deficiency occurs in many species of animals and manifests itself differently in different animals;

productivity, however, is inevitably reduced in all animals. The major clinical manifestations in animals are neonatal mortality and visible or palpable enlargement of the thyroid gland in some animals. Colloid goitre is less common in animals, however, and probably represents an evolutionary stage after primary hyperplasia. Goitre, therefore, should not be considered as the only criterion for iodine deficiency in animals as its manifestation is species dependent. The importance of subclinical iodine deficiency – the death of a newborn without goitre – is more important.

Reproductive failure is the outstanding manifestation of iodine deficiency in farm animals. Irregular or suppressed estrus causing infertility has been reported in animals in iodine deficient areas. Development of the fetus is retarded or arrested at some stage during gestation resulting in many problems before and after birth. The main point is that iodine deficiency, and not simply goitre, also influences the iodine status of the offspring. Many articles describe the effects of IDD in different species of livestock, including cattle, horses, sheep, goats, and poultry.

Productivity

Milk production is positively correlated with iodine intake for many animals. Longterm deficiencies may result in decreased milk yields that are important to offspring health as well as for human consumption. 'Global occurrence of iodine deficiency disorders in livestock' describes this aspect in detail.

Requirements of Iodine

The minimum dietary iodine requirements of farm animals are difficult to express with accuracy as the requirements for growth may not be identical with those for reproduction and lactation. Iodine requirements also vary with the type and amount of goitrogens found in the feeds and iodine supplementation is necessary. A clear consensus is hard to find amongst the different articles on this issue. However, several tables and discussions can be found throughout the book documenting suggested levels ('Global occurrence of iodine deficiency disorders in livestock', 'Iodine Deficiency Disorders and iodized salt for livestock', and 'Iodine deficiency and animal production').

Iodine Supplementation

The choice of control measures in animals is largely dependent on the severity of the problem and the economic stakes involved. Supplemental iodine may be incorporated into salt, mineral mixtures, or concentrate feeds of animals. Iodine has been found to be most efficiently delivered, however, through salt for many reasons. Other methods of iodine supplementation consider many iodine-containing compounds (elemental iodine, iodides, iodates, organic iodine preparations, seaweed meal, thyroid gland meal, etc.). Vehicles of iodine include: common salt, vegetable oils, water, feed, soil, plants and even some specifically designed porous vessels and intrauminal devices which stay in the rumen of ruminants for several months, slowly releasing iodine and other micronutrients. Many studies are quoted throughout the book on measures to control iodine deficiency in livestock and other related issues as found in 'Control measures for iodine deficiency in livestock: current status', 'Iodine Deficiency Disorders and iodized salt for livestock', 'Universal Salt Iodization programme in India', and 'Current status of IDD control programme in India'.

Iodine Toxicity

In humans, it has been reported that an intake of 2000 μg of iodine/day is potentially harmful. In all cases, however, there is no risk of toxicity to any class of animal, even with salt containing high levels of iodine of 200 ppm. 'Global occurrence of iodine deficiency disorders in livestock', 'Iodine Deficiency Disorders and iodized salt for livestock', and 'Control measures for iodine deficiency in livestock: current status' discuss this topic in detail.

Human Health and Economic Impact of IDD in Animals

From the point of view of human self-interest, it is important to prevent IDD in livestock. The majority of livestock is owned by landless, marginal and small -scale farmers who are categorized as the poorest of the poor in societies of India and other developing countries. The livestock sector provides valuable animal protein for the human diet in the form of milk, meat, eggs and chicken. It also provides draught power for agricultural operations and rural transportation as well as dung for maintaining soil fertility and fuel for cooking in rural homes. Livestock generate employment for the rural population, particularly for women of India and other developing countries. 'Role of livestock in the Indian economy' provides a good background in understanding how important the livestock sector is, to other similar developing nation economies as well.

Using the most conservative estimates, the cost-benefit ratio of IDD elimination programmes is 1:3. If benefits related to education and livestock populations are included, the ratio would be 1:8 or even more. The cost of iodization is low: normally in the range 2—7 US cents per kilogram which is less than 5 per cent of the retail price of salt in most countries. If salt for human consumption is only iodized, however, then there exists a situation in the market where non-iodized salt, which is cheaper than iodized salt, is available for animal consumption. This leads people to purchase the cheaper, non-iodized salt, especially when farmers buy one type of salt from the market for both their livestock and their family. 'Iodine Deficiency Disorders and iodized salt for livestock' and 'Current status of IDD control programme in India' discuss some of these topics.

In view of the importance of livestock in the economy, it is necessary to tackle the situation of iodine deficiency in livestock with the same determination as with humans. Research of investigators in the medical and veterinary sciences is currently being coordinated in an effort to eliminate iodine deficiency disorders in both man and livestock. The recommendations made at this meeting and a '10 - point action plan to control Iodine Deficiency Disorders in livestock' found in 'Iodine status of feeds and fodders' help to provide a concrete path towards realizing a clear message that emerges from this report: that IDD affects both animals as well as humans and that their interrelationship, especially in India, requires that attention be focused at the political level on preventing IDD in livestock. Bringing together professionals from various disciplines – the 'stakeholders' of IDD prevention – helps to catalyze action towards realizing the universal goal of creating a stable infrastructure to sustain the elimination of IDD by 2000. This is the purpose of this report.

In the compilation of this report, it was our sincere effort to ensure that every fact reported be referenced in the accompanying bibliography of each article; however, in some cases, this information could not be retrieved.

Glossary

AIIMS	All India Institute of Medical Sciences
Crore	ten million (10,000,000)
Dam	mother (of lambs, calves, etc.)
DM	dry matter
DGHS	Director General of Health Services
IDD	Iodine Deficiency Disorders
ICCIDD	International Council for Control of Iodine Deficiency Disorders
Lakh	one hundred thousand (100,000)
MOHFW	Ministry of Health and Family Welfare
MDIS	Micronutrient Database and Information System
MI	Micronutrient Initiative
MIT/DIT	monoiodotyrosine/diiodotyrosine
NRC	National Research Council
NGCP	National Goitre Control Programme

NIDDCP	National Iodine Deficiency Disorders Control Programme
NIN	National Institute of Nutrition, Hyderabad
ppm	parts per million (equivalent to milli-grams per kilogram)
T_3	triiodothyronine
T_4	thryoxine
UNICEF	United Nations Children's Fund
USI	Universal Salt Iodization
WHO	World Health Organization

PART I

Iodine Deficiency Disorders: An Overview

1

Iodine Deficiency Disorders — A Global Phenomenon

Chandrakant S. Pandav, B. R. Shamanna, K. Anand, N. K. Arora, M. G. Karmarkar

Iodine deficiency is the world's single most significant cause of preventable brain damage and mental retardation. The clinical and subclinical manifestations of iodine deficiency are collectively included in the term Iodine Deficiency Disorders (IDD) and are shown in Table 1.1.

It is now evident from detailed country surveys that the total number of people at risk of iodine deficiency exceeds 1570 million people (29% of the world's population). Globally, the prevalence of goitre is estimated to be 12 per cent; over 26 million suffer from brain damage; and 5.7 million are cretins. The highest regional goitre prevalence is in the Eastern Mediterranean Region (22.9%) with over 42.5 per cent of the population considered to be at risk of iodine deficiency in that region. In absolute terms, South-East Asia (which includes India, Bangladesh, and Indonesia) and the Western Pacific (which includes China) together account for more than 50 per cent of the world's population at risk. IDD continues to be a problem in most countries in Europe (even Western Europe, including 10 million people at risk in Germany), necessitating continuing control measures and vigilance. Table 1.2 illustrate's the number of people at risk of iodine deficiency and the number of people who suffer from goitre worldwide by the WHO and UNICEF global regions, respectively. The incredible impact of IDD is obvious from these figure and yet, it is difficult to illustrate the reality of the daily suffering experienced.

Table 1.1. Spectrum of iodine deficiency disorders

Stage in Life	Health Effects
Foetus	Abortions Stillbirths Congenital Anomalies Increased Perinatal Mortality Increased Infant Mortality Neurological Cretinism: - mental deficiency - deaf-mutism - spastic diplegia - squint Myxedematous Cretinism: - mental deficiency - dwarfism Psychomotor defects
Neonate	Neonatal goitre Neonatal hypothyroidism
Child and Adolescent	Goitre Juvenile hypothyroidism Impaired mental function Retarded physical development
Adult	Goitre with complications Hypothyroidism Impaired mental function

Source: *From Hetzel (1987), with permission.*

On an average, children living in iodine deficient areas have an Intelligence Quotient (IQ) 13 points lower than children living in iodine sufficient areas. Even though the visible impact of iodine deficiency on humans is understood, the subtle, less visible effects, such as impaired learning disabilities are also being recognized. There is yet another group, namely the livestock population, whose life and productivity is also impaired by iodine deficiency. From the point of view of human self-interest, it is important to prevent IDD in livestock, if good yields of milk, eggs, wool, and meat are to be ensured.

To prevent IDD, the body requirement is just 150 μg of iodine. Of the two most common methods of iodine supplementation — iodized salt and iodized oil injection - the former is more widely practised. The cost of salt iodization is approximately 0.05 US dollars per person per year. Using the most conservative estimates, the cost-benefit ratio of IDD elimination programmes is 1:3. If benefits related to education and livestock populations are included, the ratio would be 1:8.

Prevention, control, and eventual elimination of IDD at the country level requires long-term, sustainable action. For virtually all countries, this requires the establishment of a Universal Salt Iodization programme in which all salt for human and animal consumption is fortified with iodine. Such a comprehensive, sustainable effort requires the input of several sectors including consumer groups, salt manufacturers, social welfare groups, educators, the food processing industry, nutritionists, government, and those who work with mentally disabled children.

It has been demonstrated that if IDD elimination programmes are to be successfully sustained, they must be carefully monitored. Only then, can a country ensure that salt is adequately iodized and is reaching the entire population. Other key components of national IDD elimination programmes are education, social marketing, training, and legislation.

Of the 169 countries in the world for which data are available, 118 countries are known to have a total goitre rate of more than 5 per cent. Eighty-two countries have a salt iodization programme.

Table 1.2. Estimate of population at risk of iodine deficiency and goitre in the world by WHO and UNICEF regions

WHO Regions

WHO Region	Population (millions)	Population at risk (millions)	Population with Goitre (millions)
Africa	550	181	86
Americas	727	168	63
Eastern Mediterranean	406	173	93
Europe	847	141	97
Southeast Asia	1,355	486	176
Western Pacific	1,553	423	141
Total	5,438	1,572	655

UNICEF Regions

UNICEF Region	Population (millions)	Population at risk (millions)	Population with Goitre (millions)
East & Southern Africa	261	90	51
Central & Western Africa	271	88	34
Middle East & North Africa	338	143	71
East Asia & the Pacific	1,724	557	212
South Asia	1,183	410	149
Americas & the Caribbean	444	168	63
Developed/ Industrialized	1,217	116	76
Total	5,438	1,572	655

Source: *WHO, Global Prevalence of Iodine Deficiency Disorders, published jointly by WHO, UNICEF and ICCIDD, 1993.*

In light of the large number of individuals affected, there has been an awakening all over the world in the last ten years to understand both the magnitude of the problem and the potential ability to eliminate IDD. Almost all the nations of the world have now endorsed and supported the goal of elimination of IDD by the year 2000. The goal of elimination of IDD as a public health problem by the year 2000 was accepted by the international community in 1990 when two major decisions were taken. The first major decision was the resolution of the 43rd World Health Assembly (WHA) in Geneva in May 1990 calling for the elimination of IDD as a public health problem by the year 2000. The WHA is the annual meeting of over 180 national government health authorities which decides the policy of the World Health Organization. The second major decision was made in September 1990 by the World Summit for Children at the United Nations in New York. Both WHO and UNICEF confronted the world's leaders with a Plan of Action which included the challenge of Universal Salt Iodization as well as several other equally inexpensive methods for the future health and education of children around the world. Representatives from 159 nations have ratified and committed themselves to this plan.

To achieve the goal of IDD elimination by 2000, it was subsequently agreed that all countries would attempt the iodization of at least 95 per cent of their salt supplies by the end of 1995. Of the 94 countries with IDD problems, 58 have achieved the goal of iodization, 32 can achieve it with an accelerated effort, and the remaining 4 are unlikely to do so at the present rate of progress.

There is considerable confidence that IDD will be eliminated by 2000. This goal has already been reached by a number of developed countries and rapid progress in many developing countries is being made. The USA and Canada have eliminated IDD. Norway, Sweden, the United Kingdom, Australia, and New Zealand have also achieved this goal. More recently, at the Joint PAHO-WHO/UNICEF/ICCIDD Regional Meeting on the 'Elimination of Iodine Deficiency Disorders' in the Americas held in Quito, Ecuador in April 1994, it was announced that Ecuador has achieved the goal of maintaining iodization of 95 per cent of its salt supplies.

Bolivia has recently requested for verification of its success. Nearer home, Bhutan has also approched the target of Universal Salt Iodization.

Iodine deficiency is one of the micronutrient deficiencies which causes 'Hidden Hunger'. Speaking at 'Ending Hidden Hunger', a policy conference on micronutrient malnutrition, at Montreal (Canada) in October 1991, Mr. James P. Grant, Executive Director UNICEF said:

> "*Micronutrient deficiency does not produce hunger as we know it; it gnaws at the core of health, but not in the belly. Most of its consequences are not really perceived; like the iceberg, its bulk lies beneath the surface. Even its most apparent effects, such as blindness and cretinism, seem to most people to be unrelated to diet. That is why we call it 'hidden hunger' and why such an extraordinary effort must be made through every available channel to drag it into the open, make it visible as an issue at the political level, and empower families with the prevention knowledge they need.*"

Conclusion

Let us all come together and build a partnership so that we can successfully eliminate this ancient scourge of mankind.

References

Hetzel, B.S., C.S.,Pandav. (eds). *1994. SOS for a Billion: the Conquest of Iodine Deficiency Disorders.* Oxford University Press: Delhi.

WHO/UNICEF/ICCIDD. 1993. 'Micronutrient Deficiency Information System – Global Prevalence of Iodine Deficiency Disorders.' MDIS Working Paper #1.

Pandav, C.S., H., Viswanathan, D.P., Haxton. 1995. 'Sustaining Elimination of Iodine Deficiency Disorders'. *Nat. Med. J. India.* 8 (2): 49-51.

2

Universal Salt Iodization Programme in India

R. Prakash

Abstract

India has earnestly taken up the challenge of tackling IDD through the Universal Salt Iodization programme. Backed by the strong determination of the government, the country has made phenomenal progress since 1983 in various aspects relating to the National Iodine Deficiency Disorders Control Programme. India indigenously produces potassium iodate and fabricates iodization plants. It manufactures iodized salt that meets the full requirements of goitre endemic areas and exports it to some countries. A quality control mechanism has been developed and legislation has been introduced to strengthen the monitoring of quality control measures. India now offers its expertise to developing countries in the fabrication and installation of salt iodization plants as well as potassium iodate manufacture. A communication package through information, education and communication has been developed. It also supplies spot-test kits to determine the iodine content of salt.

Introduction

The National Goitre Control Programme (NGCP) of India was established in 1962 to survey and initiate measures necessary to eliminate and control endemic goitre. As goitre surveys progressed, it was discovered that no part of India is free from goitre and that goitre must be recognized as a national health problem. With the detection of cretinism and of mental retardation amongst the population, the magnitude of the problem is now considered to extend over a range of disorders termed as 'Iodine Deficiency Disorders' (IDD). The Government of India constituted two committees under the Chairmanship of the Salt Commissioner: a Working Group on the Salt Industry and a Study Group on Iodized Salt. On the basis of both of their recommendations, the government restructured the NGCP in 1986 by adopting the policy of Universal Salt Iodization (USI) to control IDD in India. The programme is now known as the National Iodine Deficiency Disorders Programme (NIDDCP).

Production of iodized salt increased from 0.2 million tonnes in 1983 to 3.2 million tonnes in 1992. The remarkable success of the NIDDCP is due to the following reasons:

1) Political commitment to the programme at the highest level by adoption of USI to be achieved in a phased manner and its inclusion in the National Health Policy.

2) Incorporation into the new 20-point programme of the Prime Minister in 1986 under 'Health for all by 2000 AD'.

3) Increasing the financial allocation from Rs 2 lakhs in previous plans to Rs 200 million in the 7th Five Year Plan.

4) Identifying a single nodal agency (Salt Department) for monitoring the production, distribution, and quality control of iodized salt.

5) Creation of central government-supported 'Goitre control cells' in all states / union territories to monitor the programme.

6) Legislative measures enacted by the state governments/union territories by the issue of notifications under the Prevention of Food Adulteration Act 1954 for the compulsory use of iodized salt instead of non-iodized as the common salt for edible purposes.

7) Better management by redefining the role of each agency in such a multidisciplinary and intersectoral programme.

8) Creation of public awareness through education, seminars, training programmes, and mass media (press, TV, radio, posters, etc.).

9) The private sector was permitted to manufacture and market iodized salt since 1983.

10) Subsidization and full reimbursement of the cost of potassium iodate used for the iodization of salt in the initial years.

11) Legislation that considers the loss of iodine salt in transit and mandated the level of iodine in salt to 30 ppm at the manufacturing level and 15 ppm at the consumer level.

12) Periodical seminars and workshops held at various production and distribution centres involving administrators, medical practitioners, salt producers, distributors, retailers and government organizations.

Prevalence of IDD

About 20 per cent of the nation's 843 million people are living in areas exposed to a high risk of iodine deficiency disorders. The prevalence of goitre in different states of the country is shown in fig. 2.1.

Nodal Agency

The NIDDCP is a multi-departmental organization. The Ministry of Health & Family Welfare is the policy making body of the Government of India, while the Salt Department, under the Ministry of Industry, Government of India is the nodal agency for monitoring production, distribution, and quality control of iodized salt. Each state government has a 'Goitre control cell' which carries out periodic surveys of goitre prevalence and reports to the Ministry of Health & Family Welfare. The Directorate General of Health Services under the Ministry of Health & Family Welfare co-ordinates its efforts with the state units.

The state governments are responsible for the distribution of iodized salt within the state through a public distribution system or through the open market. The Director of Civil Supplies of each state issues a notification banning the sale of non-iodized salt for edible purpose in a goitre-endemic area. The Salt Commissioner, in consultation with the Ministry of Railways, arranges for the movement of iodized salt from the production centre to the goitre-endemic area. The annual requirement of each state is worked out on the basis of the consumption pattern of six kg per annum per individual (which also includes the requirement of animals). The population growth rate is also taken into account when periodic revisions are done to assess the requirements of each state.

Current Status of Salt Iodization

Process of Iodization

The approved processes of iodization of salt in the country are:

1. Spray mixing process,
2. Submersion process, and
3. Batch process.

Fig. 2.1. Iodizing salt at the Great Sambhar Lake Salt Works in Rajasthan, India, where UNICEF provided the first iodization plant (much more massive than this one) in 1992 (photograph J.P. Greaves 1993).

Location of Iodization Plants

Iodization plants are freely allowed to be set up in the country. While installing plants at consumption centres is not discouraged, it would be uneconomical owing to the multiple stages of handling involved.

Permits

No fee is charged for issue of permit to install an iodization plant. In fact, free assistance is rendered to entrepreneurs for commissioning the plants without delay.

Source of Iodine

The entire supply of iodine of the required quality is imported and converted to potassium iodate for the iodization of salt in the country.

Specification for Iodized Salt

Iodized salt will be crystalline, solid, white, pale pink or light grey in colour. It shall not contain moisture in excess of 6 per cent of the weight of the undried sample. It shall contain on a dry weight basis:

1. At least ninety-six per cent by weight of Sodium Chloride.
2. Not more than one per cent by weight of matter insoluble in water.

3. Not more than three per cent by weight of matter soluble in water other than sodium chloride.

4. An iodine content of not less than 30 ppm at the manufacturer's level and 15 ppm at the retail level.

Packing Material

Iodized salt shall be packed in high density polyethylene/polythene lined jute bags of 50/75/100 kg capacity. It can also be packed in low density polyethylene packets of 1/0.5 kg for sale.

Legislation

All goitre-endemic areas shown in fig. 1 have been notified by the relevant state governments to ban the sale and distribution of uniodized common salt for edible purposes; legislation has been introduced for this purpose. Only iodized salt as per specifications can be sold in these areas. Under the Prevention of Food & Adulteration Act, if on analysis, iodized salt is found to be sub-standard, all persons who have manufactured, distributed, and sold this salt (right from the production to the retail levels) will be prosecuted and punished with imprisonment which may extend from six months to two years and also fined not less than Rs 1000/-, besides forfeiting that iodized salt.

Salt Production in India

Common salt is mainly produced in India from sea water along the east and west coasts and from sub-soil brine in the states of Gujarat and Rajasthan by solar evaporation in salt works. The average annual production is more than 13 million tonnes, sufficient to meet the requirement for human consumption (5 million tonnes) and industrial use, leaving a surplus for export. Salt producing states in order of production are: Gujarat, Tamil Nadu, Rajasthan, Maharashtra, Andhra Pradesh, Orissa, and West Bengal.

Primary Salt Producers

There are about 9000 salt producers in the country. The salt works are classified into four categories: above 100 acres (category I), 10 to 100 acres (category II), up to 10 acres (category III), and small individual manufacturers requiring no permission for manufacture of salt (category IV). The approximate number of units by category and percentage of contribution to total production are: category I – 450 (55%), category II – 650 (4%), category III – 1900 (10%), and category IV – 6000 (31%).

Number of Iodization Plants and Capacity

The Salt Commissioner to the Government of India has permitted 529 units to produce and market iodized salt. The details of the units and capacity are given in table 2.1 (as of 31 March 1993). The installed capacity is about 20% more than required, assuming that all edible salt in the country is to be iodized.

Table 2.1. Salt producing/marketing units by state and annual capacity

State	No. of Units	Annual Installed Capacity (lakh tonnes)
Gujarat	230	32.01
Rajasthan	184	20.01
Tamil Nadu	43	4.72
Andhra Pradesh	34	2.63
Maharashtra	12	0.85
Other states	26	5.11
Total	529	65.33

Production of Iodized Salt

The current level of production of iodized salt in the country by authorized iodization units is about 2.8 million tonnes and 0.4 million tonnes by unauthorized units. This meets the requirements of all notified areas (where goitre prevalence is high and where sale of non-iodized salt is banned). The production of iodized salt will further pick up if ban notifications are issued for the remaining area and infiltration of non-iodized common salt from non-notified areas is checked by state governments.

Distribution

Rail transport plays a vital role in distribution of iodized salt to various consuming states. The salt/

iodized salt moves by rail and road from the three major salt producing states of Gujarat, Rajasthan, and Tamil Nadu to the various consuming states. Distribution of iodized salt by rail is governed by the zonal scheme, whereby the salt producing states are linked to particular states of the country. This zonal scheme is framed by the Salt Department in consultation with the railway authorities. It ensures a reliable supply of this commodity to every nook and corner of the country throughout the year. The states have managers to monitor the distribution of salt through secondary channels to various wholesale depots, stockists, and retail outlets. State governments also have an important role to play, in that they must monitor the state demand for salt based on the population and yearly consumption of six kg of salt per individual. The distribution of salt in the north-eastern states and West Bengal is controlled by the nominee system, whereby these state governments appoint certain nominees to lift the iodized salt from the salt production centres. The nominee system, however, does not exist in other states of the Union where any trader is free to lift salt to the states.

Quality Control

At Production Level

Every iodized salt manufacturer has to maintain a process control laboratory to ensure proper level of iodization. Samples are drawn during the manufacture of iodized salt and are analysed. Whenever there is less or excess iodine in the salt, corrective action is initiated by the salt manufacturer. A register of such analysis is maintained by the producer.

A second check is carried out by the Salt Department before the despatch of salt by rail or road. Samples are drawn at random and analysed in the departmental quality control laboratories. Whenever any sample is found to be sub-standard, the manufacturer concerned is asked to reprocess that batch of salt. Permits are suspended/cancelled in the case of repeated failures.

At Distribution Channel

Quality control at the retail level is conducted by the state governments through the Health Directorates. The Food Inspectors of the concerned state governments maintain the quality of iodized salt. If the samples are found to be sub-standard, all persons who have manufactured, distributed, and sold the sub-standard iodized salt will be prosecuted under the Prevention of Food Adulteration Act.

Price

The cost of salt iodization works out to approximately Rs 50/- per tonne (i.e. 5 paise per kg). The retail price of iodized and common salt of the same quality and similar packaging are more or less the same in almost all consuming centres.

About 15 to 20 per cent of edible salt is being sold in polythene packets, especially in urban and semi-urban areas, at a cost of Rs 3/- to Rs 3.50/- per kg. The retail price of loose salt of similar quality is approximately half the cost of packed salt. In rural areas, salt is sold mostly in loose form.

Incentives

Iodized salt producers who have set up plants with specific permission from the Salt Department are eligible for the following incentives:

1. Priority in rail movement: Transport of iodized salt from production to consuming centres has been accorded a high priority. This will enable quick and efficient movement.

2. Market: Large areas of the country are progressively being notified by the respective state governments to receive only iodized salt. This will increase and guarantee the market for iodized salt.

3. Technical Assistance: The Salt Department gives free technical guidance in setting up iodization plants, training of personnel in operation and maintenance of plants, and in quality control.

4. Financial Assistance: Industrial Development Bank of India extends financial assistance to the private sector through commercial banks for setting up iodization plants.

5. Quality Control: Iodized salt producers can have samples tested in the Salt Department's quality control laboratories free of charge.

6. Training Courses: The Salt Department holds training courses periodically for the benefit of laboratory technicians.

7. Exports: Export of iodized salt is exempted from payment of cess. Export of iodised salt is accorded a higher priority 'B' within the same category of movement.

Problems in Programme Implementation

Technical Problems

i) Non-issue of ban notification

The progress in iodized salt production is directly proportional to the number of notified areas. Although the capacity permitted is much more than the target fixed, the production of iodized salt remains stagnant at about 3.2 million tonnes. States governments that still have not issued ban notifications are constantly being persuaded to realize the importance of issuing such notifications.

ii) Difficulties in setting up salt iodization plants in small scale sectors

Iodized salt is manufactured mainly by two processes: spray mixing and submersion. The cost of a spray unit varies from Rs 1.5 to Rs 4 lakhs depending on the quality of material used in its fabrication. The installation cost of a submersion unit is around Rs 40,000 to Rs 50,000. A batch-type plant has also been developed at a cost slightly more than Rs 50,000 depending on the capacity and compressor used.

As the submersion plant is labour-oriented, its operational cost is comparatively high and is thus not economically suited for salt manufacturers in the medium and small-scale sector. The conventional type spray plants are costly to set up. The Salt Department is facing a lot of constraints in tackling these problems of the medium and small-scale salt manufacturers, which number more than 6000. Efforts are nonetheless, being made to ensure proper iodization at this scale of production.

iii) Baragara salt – difficult to iodize uniformly

Baragara variety of salt consists of big crystals and is produced from the sub-soil brine in Little Rann and Kutch, Gujarat State. It contributes about 1/6 of the country's total salt production. Baragara salt is mostly despatched to Uttar Pradesh as well as Parts of Madhya Pradesh, Rajasthan and Delhi. The size of the crystals poses a problem for iodization. It is also difficult to pack baragara salt before use which results in the reduction of iodine content of salt. Steps have been initiated to prevail upon the baragara salt manufacturers to produce smaller crystals or to crush baragara salt before iodization; to create awareness in the minds of consumers to demand only crushed iodized salt; and to supply crushers with UNICEF assistance.

Railway Constraints

Railways, which is the cheapest mode of inland transport, is the chief carrier of iodized salt. Much depends on its ability to transport iodized salt with minimal loss of iodine content across long distances. Inadequate/irregular placement of wagons affects the timely supply of quality iodized salt to consumers.

Administrative Constraints

Adequate staff strength is lacking to run the National Iodine Deficiency Disorders Programme successfully.

Financial Constraints

Delay in the allotment of a sufficient budget grant often causes set backs to the programme. It also affects publicity programmes designed to create awareness, especially among villagers at the grassroots level.

UNICEF Support

UNICEF plays an active role in the production of iodized salt and in quality control monitoring. It has donated 17 iodization plants which have been installed in both salt producing and salt consuming states. It also provides assistance to orientation and training programmes, helps establish monitoring laboratories at the production level, and supports the Salt Commissioner's office in organizing workshops of

IDD, situational analysis on supplies and availability of iodized salt in selected districts in India, and in the development of field testing kits. In 1988, the Government of India with UNICEF support initiated an information and education campaign to strengthen the supply and demand of iodized salt. A professional advertising agency was commissioned for the task which resulted in a communications strategy on IDD that was supported by the results of a knowledge, attitude, and practice (KAP) study. Project document for IDD control in India, a UNICEF supported initiative, has been approved for 1993-5 and efforts to achieve the targets set out in the document continue.

Management of Iodised Salt Production and Distribution

From the preceding discussion, it can be seen that the Salt Department, which is the nodal agency for the implementation of NIDDCP in India, maintains a close liaison with all of the concerned agencies. Despite various constraints, the Salt Department has succeeded in ensuring the supply of fully satisfactory iodized salt to all goitre notified areas and has the capacity to meet the needs of the entire country.

Current Status of IDD Control Programme in India

N. Pralhad Rao and G. N. V. Brahmam

Introduction

Goitre is one of the many, but most obvious manifestations of Iodine Deficiency Disorders (IDD). As an epidemiological indicator, goitre has been widely used to measure the magnitude and endemicity of IDD. Its prevalence in endemic form in the sub-Himalayan regions of the country (extending from Kashmir in the north west to Assam, Arunachal Pradesh and Nagaland in the east) has long been known. The districts lying in this at-risk belt have an estimated population of 150 million, of which nearly 55 million are likely to suffer from goitre, 6 million from neurological defects, and 2.2 million from cretinism. The Kangra Valley Study, pioneered by Prof. Ramalingaswami and his colleagues at the All India Institute of Medial Sciences (AIIMS), showed for the first time in India that the distribution of iodized salt drastically reduces the prevalence of goitre. Encouraged by these results, the Government of India initiated the National Goitre Control Programme (NGCP) in 1962 with the following objectives (GOI, 1962; Sooch, 1973; Hetzel, B.S., 1991; Pandav, 1994):

1. To identify and assess the prevalence of goitre in suspected areas through systematic surveys.

2. To prevent and control goitre through the production and supply of iodized salt in place of common salt in the endemic areas.

3. To assess the impact of the programme by resurveying areas after five years of continuous supply of iodized salt.

Realizing the wider implications of the effects of iodine deficiency on growth, brain development, and mental function, the programme was redesignated in April 1992 as the National Iodine Deficiency Disorders Control Programme (NIDDCP) (Hetzel, 1991; GOI, 1992)

Surveys conducted by several agencies, such as: the central IDD control team of the Directorate General of Health Services (DGHS), Ministry of Health, All India Institute of Medical Sciences (AIIMS), National Institute of Nutrition (NIN), and the Indian Council of Medical Research (ICMR), have revealed that innumerable pockets of endemic goitre exist in regions other than the sub-Himalayan parts of the country. The extra-Himalayan endemic foci include: peri- and sub-Vindhya areas of Madhya Pradesh, Aravali ranges in Rajasthan, Narmada valley in Gujarat, Aurangabad, Wardha and Pune districts in Maharashtra, tribal areas of Visakhapatnam and Adilabad districts in Andhra Pradesh, and tea estates in Kerala. Recent figures from the Directorate General Health Services suggest that about 82 per cent of those districts (197 out of 239) surveyed for goitre have been found to be endemic. It

Table 3.1. Prevalence of IDD in different states & Union Territories

No.	States/UT	Number of Districts		
		Total	Surveyed	IDD Endemic
1	Andhra Pradesh	23	7	6
2	Arunachal Pradesh	12	10	10
3	Assam	18	18	18
4	Bihar	38	21	20
5	Goa	1	1	1
6	Gujarat	19	11	9
7	Haryana	14	3	2
8	Himachal Pradesh	12	10	10
9	Jammu & Kashmir	15	14	14
10	Karnataka	20	17	5
11	Kerala	14	5	4
12	Madhya Pradesh	45	16	16
13	Maharashtra	31	28	10
14	Mizoram	4	3	3
15	Manipur	8	6	6
16	Meghalaya	5	2	2
17	Nagaland	7	7	7
18	Orissa	13	2	2
19	Punjab	12	3	3
20	Rajasthan	27	3	3
21	Sikkim	4	4	4
22	Tamil Nadu	21	2	1
23	Tripura	3	3	3
24	Uttar Pradesh	62	34	29
25	West Bengal	18	5	5

Table 3.1. (Contd)

No.	States/UT	Number of Districts		
		Total	Surveyed	IDD Endemic
1	Andaman and Nicobar	2	1	1
2	Chandigarh	1	1	1
3	Dadra & Nagra Haveli	1	1	1
4	Daman & Diu	1	1	1
5	Delhi	1	1	1
6	Lakshadweep	1	--	--
7	Pondicherry	4	--	--

Source: *Sheila Vir. Control of Iodine Deficiency. The National Programme – Current Status. NFI, 15, 2nd April, 1994.*

should be noted, however, that only 52.2 per cent of all districts (239 out of a total of 457) were surveyed for this purpose (table 3.1). Thus, the problem of goitre is definitely more widespread and its severity more intense than what was earlier believed. The control programme of iodized salt distribution implemented over the past three decades has failed to produce the desired effect due to many constraining factors.

Constraints

The factors responsible for the failure of the programme, as reported in a review by the Nutrition Foundation of India (NFI) (Gopalan, 1981), were more operational and administrative than technological. The main issues, as have been observed in other Nutrition Action Programmes, was non-availability of the basic input (nutrients) in adequate quantity to match the requirement, frequent interruptions in the supply line, and ineffective distribution (Pralhad Rao, 1990). Production and procurement of iodized salt has fallen short of the requirement. This factor was further compounded by other factors, such as: lack of mechanisms to effectively coordinate, monitor, and enforce legal provisions of the Prevention of Food Adulteration (PFA) Act, lack of facilities to check salt for iodine content (quality control) along the distribution line, and lack of awareness in the public as well as in the implementing agencies about the health and economic implications of iodine deficiency.

Inadequate Production

The estimated annual production of common salt in the country is around 13 million tonnes. Sub-soil brine in Gujarat and Rann of Kutch and sea-water in the coastal areas are the main sources of salt. The most common method employed in its production is solar evaporation. Gujarat, Tamil Nadu, and Rajasthan are the main producing centres. Nearly half of the total production of salt goes to human and animal consumption and the rest is used in industry, export and buffer stocking. There are about 10,000 salt manufacturers in the country. The majority of them are small-scale producers supplying the bulk (nearly 60%) of the edible salt required. Out of the six hundred odd licensed salt manufacturers, about 550 are registered producers of iodized salt.

During the period from 1978 to 1983, the production of iodized salt from plants in Rajasthan, Gujarat, and West Bengal (established with UNICEF support) did not exceed 1.3 lakh (0.13 million) tonnes. The disappointing feature of production was not that this output was much below the requirement of iodized salt in endemic areas, but that it was quite low compared to the plant installed production capacity of 3 million tonnes production (table 3.2). In the latter half of the 1980s, under the liberalized policy of Universal Salt Iodization adopted by the government, the production of iodised salt increased from about 2 lakh tonnes in 1985-6 to about 14 lakh tonnes in 1987 and 23.8 lakh tonnes in 1989-90 — a dramatic increase of several hundred fold. The latest figure of iodized salt production from the Salt Commissioner is around 29 lakh tonnes.

Table 3.2. Production, production capacity, and requirement of iodated salt (thousands of tonnes)

Year	Production of Iodated Salt	Production Capacity	Iodated Salt Requirement
1978	106	--	--
1979	120	--	--
1980	122	--	--
1981	129	376	1,000
1982	123	--	--
1983	130	--	--

Source: *Iodine Deficiency Disorders in South East Asia. SEARO. Regional Health Paper No. 10. WHO: New Delhi. 1985. p. 44.*

Table 3.3. Retail selling price of iodized salt in various states (January to April, 1989)

State	Rate (Rs/kg)	
	Crushed (Loose)	Powdered & Packed in Polythene Packets
Uttar Pradesh	0.60 - 1.00	1.50 - 2.00
Bihar	0.80 - 1.00	1.00 - 2.00
Assam	1.00 - 1.20	1.50 - 2.00
Sikkim	1.10	---

Source: *Salt Commissioner, Govt. of India, (Background paper at the Expert Group Meeting on IDD Control held at the ICMR, New Delhi, 20 March, 1989.*

The production of 52 lakh tonnes of iodized salt, representing the total requirement of the country, is expected to be achieved by the year 1995.

Cost of Production

The economics of production and procurement of iodized salt suggest that production cost has never been very high. The cost of potassium iodate (KIO_3), which needs to be imported from Japan, Indonesia, the Netherlands, etc. is Rs 20/tonne; processed KIO_3 costs Rs 25/tonne. One kilogram of potassium-iodate is required for iodizing 20 tonnes of salt at a 50 ppm level of iodine. This cost works out to only 2 per cent of the total cost. Other costs, however, such as: handling, packaging in high density polyethylene (HDPE) bags, transporting, unregulated selling price, etc. make the commodity fairly expensive. Movement by road is more expensive than by rail but seems to be more convenient for traders. The distance between production points and different consuming states range from less than 600 km to over 2000 km.

Table 3.3 demonstrates the cost of iodized salt in loose and packed forms for different areas as reported by the Salt Commissioner. In the city of Hyderabad, the retail price of powdered and packed iodized salt is around Rs 4 per kg as compared to Rs 1.50 per kg for plain uniodized salt, illustrating the disparity in price.

Such a difference in cost is undoubtedly a definite deterrent to consumer demand, especially in rural and tribal areas where the purchasing power of the people is very low and the awareness of the health implications of using non-iodized salt in non-existent. Given the slow pace at which the information, education, and communication (IEC) network and the monitoring system (MIS) function within the programme framework, the need to keep the cost of iodized salt low enough to enable the common man to purchase it on a regular basis, through the Public Distribution Scheme network, is urgent.

Operational Difficulties

Apart from the production and cost factors of iodized salt, the following operational constraints have added to the difficulties of its distribution.

Transport bottlenecks

Railways account for the transportation of only about 0.5 million tonnes of iodized salt in the country. The bulk of salt is moved by road transport. Apart from steps taken by the government to allocate appropriate railway wagons on a priority basis (B priority), there is no known or regulated movement plan of iodized salt within and outside salt producing states. Unless the total distribution system through rail, road, and other means is streamlined and coordinated with production programmes, proper distribution cannot take place which in turn limits public demand for iodized salt.

Ineffective enforcement of PFA Act

The Prevention of Food Adulteration (PFA) Act has not been enforced by state health departments. This has resulted in the entry of non-iodized salt into goitre endemic areas leading to a dual market (supply) in which iodized salt is costlier than non-iodized salt. Lack of adequate knowledge about the provisions of the Act by the district civil supply authorities has led to ineffective control over the salt nominees, resulting in poor lifting and faulty storage/distribution of iodized salt. The poor quality of gunny bags used for storing salt by the traders often results in loss of iodine from the fortified salt.

Delay in ban notification

Ban notification (which should immediately follow declaration of an endemic area) is a prerequisite to the supply of iodized salt and has often been unnecessarily held up.

Lack of awareness, supervision and quality control facility

Stark ignorance about the underlying cause of IDD and the virtual lack of continuous mass education programmes highlighting the serious and irreversible physical, mental, and neurological effects of iodine deficiency are responsible for the apathy of the people towards the programme.

Absence of goitre control cells in some of the state health directorates of endemic states has invariably led to a lack of supervision of the programme. It has been reported that there are no IDD cells in seven states, let alone quality control laboratories. This results in either no co-operation or unsatisfactory co-operation by state authorities with the central teams during assessment and evaluation surveys.

New Initiatives

Realizing that operational constraints can be overcome through proper application of existing administrative procedures and that the country has enough technical know-how and resources to produce an adequate amount of iodized salt to meet the country's requirement, the Government of India has renewed it's commitment to the national IDD control programme. The objectives of the new initiative remain essentially the same as those of the NGCP. The approach in achieving these objectives, however, has been redefined and policy decisions have been altered. Salient features of the initiatives are listed below.

Universal salt iodisation

An important decision taken by the Government is to iodize all salt used for human consumption in the country (Universal Salt Iodization). In pursuance of this decision, the IDD control programme has been included in the current five year plan and special allocation has been made. The programme formed an important item on the agenda of the Prime Ministers 20-point programme. To obtain maximum production of iodized salt in the shortest possible time, permission was granted to private salt manufacturers to iodize salt. Every assistance, financial and technical, was provided to small entrepreneurs to set up low cost, low capacity iodization plants. Training of operators and technicians in conducting quality checks (quality control) was ensured. There is also a subsidy programme which allows a free supply of potassium iodate to manufacturers.

High priority for transportation of iodized salt

To ensure an uninterrupted supply of iodized salt, high priority (B category, next only to defence) has been accorded in the allocation of railway wagons for transportation of salt.

IEC campaigns

Greater emphasis has been placed on multimedia IEC campaigns through meetings, conferences, advertisements, workshops, etc. to sensitize and inform the public, as well as manufacturers, traders, and functionaries at all levels of government and non-government sectors about IDD.

Organization

In order to streamline administration of the programme under the Ministry of Health and Family Welfare, which is the policy making body of the Government of India, the Salt Department under the Ministry of Industry has been chosen to act as the nodal agency for monitoring production, distribution, and quality control of iodized salt. The Salt Department is also responsible for the payment of subsidies to the manufacturers.

Each State Government, through its goitre control cell, is expected to conduct goitre prevalence surveys and provide feedback to the Ministry of Health and Family Welfare. The Ministry of Health through the Directorate General of Health Services (DGHS) is responsible for coordinating the activities of state level goitre control cells.

The ICMR and AIIMS have their own separate units dealing with special studies on IDD.

The distribution of iodized salt within the state is the responsibility of State Governments and is undertaken either through Public Distribution System (PDS) or through the open market.

The Director of civil supplies of each state is made responsible for issuing notifications banning non-iodized salt (for edible purpose) in the goitre endemic areas.

To date, a total of 25 IDD cells have been set up in different states/union territories and 23 areas (accounting for about 65% population), have been covered under the ban order.

Specification

Under PFA (Prevention of Food Adulteration Act,) it is now specified that salt should contain not less than 30 ppm iodine at the production point and 15 ppm iodine at the retail level.

Future Strategy

In the context of IDD, it should be remembered that the advocacy of increasing the intake of nutrients through diversification of diet as recommended with the other nutritional deficiency diseases, does not hold well because the problem of IDD is a problem of 'poverty of soil' rather than the 'poverty of people'. Supply of iodine through fortified foods or in medicinal form is the only measure available for its control. Hence, the main thrust of a future strategy for the IDD control programme includes the following aspects.

Increased production

The future strategy of the NIDDCP is geared to cover the entire country in a phased manner. Therefore, increasing production through spray mixing, submersion, and batch process methods, has been recommended. Special emphasis has been laid on designing manually operated least-cost iodization plants. Decentralized production can be expected to have better and wider outreach.

Effective monitoring and information system

An effective machinery of well-knit technical and administrative personnel empowered to check salt for iodine content (quality control) at production, storage, and consumption points should be made functional so that the provisions of the PFA can be enforced. In this context, establishment of IDD cells and quality control laboratories in each state and their networking should receive attention.

Increased public awareness

The population belonging to the lower socio-economic strata of society are unaware of IDD and the preventive properties of iodized salt and so prefer or are tempted to purchase plain salt because of its low cost. Consequently, an appropriate Information, Education and Communication (IEC) programme with simple messages in terms of do's and don'ts using mass media, folklore, TV, radio, etc. should be undertaken. The IEC programme should motivate people and eventually convert the present supply-oriented strategy into a demand-based programme.

Well-trained functionaries

Functionaries of the Departments of health, industry (salt), and transport (railways) should be periodically exposed to inservice training oriented to equip them with the necessary knowledge and skills needed to effectively organize/manage the programme. The training should also include knowledge about the legal provisions under the PFA Act of 1954, amended in 1964, 1976, 1981, and 1994, cover standards for edible salt, iodized salt, iron fortified salt, and potassium iodate. Under the provisions of the Act, a licence is essential for manufacturing but not for the sale and distribution of salt. Addition of new chemicals requires approval of the Central Committee for Food Standards (CCFS).

The loop holes in the rules, such as: lack of legal provisions to prevent entry of non-iodized salt in notified endemic areas (since it is allowed for industrial and animal use) pose problems of checking the quality and preventing dual supply of salt. These should be corrected by effecting appropriate amendments to the Act.

In keeping with the concept of Essential National Health Research (ENHR) advocated by the Commission on Health Research Development (Report, 1990) exploration into the possibility of extracting iodine from seaweeds grown in India and from residues in oil wells, and considering other technologies, such as iodized oil for hyperendemic areas with widespread cretinism should receive priority. The strategy of long-acting iodine prophylaxis as a one-time measure for target groups, for instance, newly married adolescent girls followed by salt prophylaxis can be expected to halt the relentless march of cretins and cretinoids in communities residing in hyperendemic areas. (It is a cruel irony that there exists so much knowledge and experience about long acting prophylaxis in neighbouring countries of Asia and yet, a decision to act even in hyperendemic states, like Sikkim and Manipur which have a high prevalence of crippling cretinism (3 - 6%) is eluding.) Extraneous (non-health) arguments, such as: 'This would encourage other states to raise the same demand' or 'This would entail foreign exchange', etc. should not be allowed to detract from preventing the irreversible forms of IDD. There is an urgent need to plan and conduct a field trial to demonstrate the efficacy of iodized oil administration in adolescent girls in preventing the incidence of cretinism and cretinoids in hyperendemic districts of the Indian subcontinent so that any controversy to this approach can be resolved.

Research agenda

Items that should form the research agenda of national institutes and other related agencies under ENHR in the context of IDD are as follows:

* A search for an indigenous source of oil for use as a base in iodized oil and its production so as to meet the requirements of hyperendemic areas.
* Assessing the role of a stabilizer, which while improving the shelf life of iodized salt, is expected to reduce the amount of iodine added to salt, thus making it less expensive.

References

Gopalan, C., 1981. 'The National Goitre Control Programme – A Sad Story'.

Hetzel, B.S., 1991. *The Story of Iodine Deficiency – An International Challenge in Nutrition.* Oxford/Delhi: Oxford University Press.

'Health Research: Essential Link to Equity in Development' 1990. Report of the Commission on Health Research for Development. Oxford University Press.

Ministry of Health and Family Welfare, Government of India, New Delhi National Goitre Control Programme, 1962.

Ministry of Health and Family Welfare, Government of India, New Delhi, National Iodine Deficiency Disorders Control Programme, 1992.

Pandav, C.S., 1994. 'IDD in South East Asia'. In: Hetzel, B.S., C.S., Pandav (eds). *S.O.S. for A Billion: the Conquest of Iodine Deficiency Disorders.* pp. 220-3 Oxford University Press: Delhi.

Pralhad Rao, N. 1990. 'Constraints in National Programmes'. Indian Council of Child Welfare (ICCW), News Bulletin XXXVIII, 4, (October - December, 1990). pp. 25-9.

Sooch, S.S., M.G., Deo, M.G., Karmakar, N., Kochupillai, K., Ramachandran, V., Ramalingaswami. 1973. 'Prevention of endemic goitre with iodized salt'. Bull. Wld. Hlth. Org. 49: 307-12.

The Salt Department, Ministry of Industry. 1994. *Universalization of Access to Iodised Salt - A Mid-Decade Goal.*

WHO. 1985. 'Iodine Deficiency Disorders in South East Asia'. SEARO Regional Health Paper, No.10. WHO: New Delhi. p 44.

4

Towards the Elimination of Iodine Deficiency Disorders in India

C.S. Pandav and K. Anand

Abstract

Iodine deficiency disorders (IDD) are an important public health problem in India with an estimated 270 million people at risk of IDD. India has adopted the strategy of salt iodization, for control of IDD and has the goal of 'Universal iodization of salt by 1995 and elimination of IDD by 2000'. There is a high degree of political commitment which needs to continue if the goal is to be achieved. Currently, the ban on sale of uniodized salt is only applicable to salt for human consumption. There is a need for extending the ban to include salt for animal consumption as IDD affects livestock as well. India has the installed capacity to produce its requirement of 5 million tonnes of iodized salt. Communication strategies have to be strengthened especially to educate people who have concerns about iodine toxicity. The success, to a large extent, depends on the quality control and monitoring of iodine content of salt at all stages from production to consumption. The NGOs and the community have to be encouraged to participate in this process. To sustain the elimination of IDD, a partnership of various stakeholders in IDD elimination is essential.

This chapter is reproduced from the paper published in the Indian Journal of Pediatrics, 62:545-55, 1995.

This paper describes the current progress towards the elimination of Iodine Deficiency Disorders (IDD) in India. It is covered under subheadings which trace the path of 'assessment - analysis - action - assessment' cycle. Under each subheading, first, the current status is described followed by the agenda for the future. Unlike communicable diseases like smallpox, IDD cannot be eradicated once and forever. The control efforts like salt iodization will have to be sustained perpetually. The concluding part of the paper raises the issues of sustainability and the partnership that needs to be built between the stakeholders to achieve the sustained elimination of IDD.

Assessment of the Problem

Current status

India is the second most populous country in the world with a population of 834 million (1991 census). High prevalence of goitre and cretinism exists in a broad Himalayan and sub-Himalayan goitre-belt from Jammu and Kashmir in the North to Arunachal Pradesh in the East and along this entire length extending at least 500 km south of the Himalayas into the flat sub-Himalayan terai (plains). In addition to the well-known 'Himalayan endemic belt', iodine deficiency has been reported from many other states in the country. In 1989, the Indian Council of Medical Research (ICMR) carried out a multicentric IDD prevalence study. Nine states outside the traditional 'goitre-belt' were studied for the prevalence of goitre and cretinism. A total of 409, 923 individuals were examined and the overall goitre

Table 4.1. Status of IDD endemicity and ban notification in states/UTs of India

State/UT	Status of ban	Total dist.	Dist. surveyed	Dist. endemic	% dist. endemic
Andhra Pradesh	Full	23	7	6	85.7
Arunachal Pradesh	Full	12	10	10	100
A & N Islands*	Full	2	1	1	100
Assam	Full	18	18	18	100
Bihar	Full	38	21	20	95
Delhi	Full	1	1	1	100
Goa	No ban	1	1	1	100
Gujarat	Full	19	11	9	82
Haryana	Full	14	3	2	67
Himachal Pradesh	Full	12	10	10	100
Jammu & Kashmir	Full	15	14	14	100
Karnataka	Partial	20	17	5	29
Kerala	No ban	14	5	4	80
Madhya Pradesh	Full	45	16	16	100
Maharashtra	Partial	31	28	10	36
Manipur	Full	8	6	6	100
Meghalaya	Full	5	2	2	100
Mizoram	Full	4	3	3	100
Nagland	Full	7	7	7	100
Orissa	Partial	13	2	2	100
Punjab	Full	12	3	3	100
Rajasthan	Full	27	3	3	100
Sikkim	Full	4	4	4	100
Tamil Nadu	Full	21	2	1	50
Tripura	Full	3	3	3	100
Uttar Pradesh	Full	62	34	29	85
West Bengal	Full	18	5	5	100
Chandigarh	Full	1	1	1	100
Dadra & Nagar Haveli	Full	1	1	1	100
Daman and Diu	Full	1	1	1	100
Lakshadweep	Full	1	-	-	-
Pondicherry	No ban	4	-	-	-
Total No. of Dist.		457	239	197	

Source: *Ministry of Industry Banning sale of edible non-iodised salt - An urgent measure. Produced by The Salt Department, Ministry of Industry with support from UNICEF, 1995.*

prevalence observed was 21.1 per cent, whereas the overall cretinism prevalence was 0.7 per cent. In India, it is estimated that 270 million people are at risk and 79 million are affected by IDD which includes, 2.2 million and 6.6 million with mild neurological disorders. Out of the total 457 districts, 239 districts have been surveyed by the Government of India and the state governments. Of these, 197 districts are endemic for goitre (table 4.1). No single state in the country is free from IDD as a public health problem on the basis of these surveys. Thus, the strategy for IDD elimination should be applied to the whole country.

The incidence of neonatal hypothyroidism (NH) in a severely endemic area of India (like Uttar Pradesh) is 2400 per 100,000 births which is 50 to 200 times more than the reported average incidence of NH in non-endemic areas. This means that on an average, with every passing hour, 10 children are born in the country who will not attain their optimal mental and physical potential as a result of neonatal hypothyroidism due to environmental iodine deficiency. It has also been observed that the Intelligence Quotient (IQ) of school children in areas with iodine deficiency is about 13 points lower than that of the children living in iodine sufficient areas. The total IQ points thus lost to the country is formidable.

All this makes India one of the major endemic iodine deficiency countries in the world. Immediate steps are therefore, required to ensure that iodine supplementation, in some form, reaches the population, at the earliest.

Agenda for the future

As there is great variability among the different districts in the country, it may not be right to extrapolate the prevalence rates to the rest of the country. It is essential that all the districts are surveyed at the earliest using epidemiologically sound techniques. The recently recommended use of full (EPI) cluster sampling technique for this purpose by a consultative committee of WHO/UNICEF/ICCIDD is easy to carry out and can be done in a short span of time. This committee also recommended the use of a simplified classification of goitre and indicators for assessing the severity of IDD as public health problem. These have been shown in tables 4.2 and 4.3 respectively. It is vital that a complete survey of the country is undertaken using this method at the earliest. However, this should not delay the immediate implementation of salt iodisation programme in these areas.

Table 4.2. Grading of goitre by the palpation method

Grade	Description
0	No palpable or visible goitre.
1	A mass in the neck that is consistent with an enlarged thyroid that is palpable but not visible when the neck is in the normal position. It moves upwards in the neck as the subject swallows. Nodular alteration(s) can occur even when the thyroid is not enlarged.
2	A swelling in the neck that is visible when the neck is in the normal position and is consistent with an enlarged thyroid when the neck is palpated.

Source: *From the report of a Joint WHO/UNICEF/ICCIDD Consultation on 'Indicators for Assessing Iodine Deficiency Disorders and their Control through salt iodization' Geneva, November 1992. Document WHO/NUT/ 94.6*

Table 4.3. Summary of IDD prevalence indicators and criteria for a significant public health problem

		Severity of public health problem (prevalence)		
Indicator	Target population	Mild	Moderate	Severe
Goitre Grade > 0	SAC[a]	5.0-19.9%	20.0-29.9%	≥ 30%
Thyroid volume > 97th percentile by ultrasound[b]	SAC	5.0-19.9%	20.0-29.9%	≥ 30%
Median urinary iodine level (μg/l)	SAC	50-90	20-49	< 20
TSH > 5 μU/1 whole blood	neonates	3.0-19.9%	20.0-39.9%	≥ 40%
Median Tg (ng/ml serum)[c]	C/A[d]	10.0-19.9	20.0-39.9	≥ 40

[a] SAC = school aged children

[b] Normal thyroid volume size values will be available from WHO and ICCIDD in 1995

[c] Different assays may have different normal ranges

[d] C/A = children and adults

Source: *From the report of a Joint WHO/UNICEF/ICCIDD Consultation on 'Indicators for Assessing Iodine Deficiency Disorders and there Control through Salt Iodization' Geneva, November 1992. Document WHO/NUT/ 94.6.*

Strategies Available

Current status

All the methods for the prevention of IDD are based on supplementation of iodine in different ways. This includes fortification of different food stuffs like salt, bread, milks, etc. with iodine or use of iodized oil. Of the two most common methods of iodine supplementation ,iodized salt and iodized oil injection, the former is widely practised. This was one of the strategies adopted by India when it launched the National Goitre Control Programme (NGCP) in 1962.

Agenda for the future

Though iodized salt is undoubtedly the strategy for the country, there is still scope for a complementary strategy in the form of iodized oil. This should, however, be limited to high endemic areas where there is high prevalence of cretinism. Other factors which further compound the problem in these areas are the distance from the salt producing centres, difficulty in transport and communication due to the hilly terrain and annual or semi-annual salt purchasing habits of people. The state of Sikkim with a high prevalence of goitre (56.9%) and cretinism (2.9%) among school children is one such example. It should, however, be clear that the iodized oil should only be a short-term strategy and all efforts to strengthen the supply of iodized salt should be made. Fortification of salt has also been tried for controlling iron deficiency anemia. With the availability of technology for double fortification, it is likely that this strategy may be adopted in the near future. This has important implications for the quality of common salt produced in the country.

Political Commitment

Current status

The efforts of the Government of India to achieve universal iodization have been affirmed at many important national and international occassions: (i) In May 1990, the 43rd World Health Assembly passed a resolution commending the governments, inter-governmental and non-governmental organizations for their efforts to prevent and control IDD as a major health problem in all the countries by the year 2000. (ii) In September 1990, the historic World Summit for Children, convened by the United Nations, adopted a plan of action setting out the goals one of which was the 'Virtual Elimination of IDD'. India was one of the signatories of the summit. (iii) Historically, goitre and IDD elimination was a part of the 20-point programme of the Prime Minister of India, late Mrs. Gandhi. Adoption of the resolution at the international Conference of Nutrition held in 1992 in Rome and the National Nutritional Policy, both of which have reference to IDD by the government in 1993, reflect the strengthening of the government's resolve to eliminate IDD. (iv) The commitment of the Government of India to universalize the iodization of salt by 1995 and elimination of IDD by the year 2000 was reflected in the second SAARC Conference on children in South Asia, held in Colombo, Sri Lanka, in September 1992. (v) The National Plan of Action for Children (NPA) of the Government of India also includes the 'Universal iodization of salt for the reduction of IDD to 10 per cent by 2000'.

Agenda for the future

In 1996, a report has to be placed before the World Health Assembly on the progress achieved in the global efforts towards IDD elimination. India can seize this opportunity to move/sponsor the resolution in view of its significant achievements.

Advocacy

Current status

Though there has been sufficient political commitment for the elimination of IDD in India, there is a need for different national and state level discussions. A meeting of the 'Policy and advocacy on IDD' was held for high level health officials on 17 January 1994 at Delhi. It was chaired by the Health Secretary and 47 officials from the central government, state governments and representatives from national and international agencies participated in the meeting.

Agenda for the future

There is a need to continue the exercise on an ongoing basis as the political and the administrative leadership keeps on changing. The advocacy efforts should focus on the potential aspect for success of this programme among the different national health programmes.

Legislation

Current status

The sale of uniodized salt has been banned by issuing notifications under Section 7 (Clause iv) of the Prevention of Food Adulteration (PFA) Act by the state governments within their territories either fully or partially. The qualities of iodized salt have been

Table 4.4. Criteria for tracking progress towards the World Health Assembly/World Summit goals: elimination of IDD as a public health problem

Indicator	Goal
1. Salt iodization	
Proportion of households consuming effectively iodized salt	>90%
2. Urinary iodine	
Proportion below 100 μg/l	<50%
Proportion below 50 μg/l	<20%
3. Thyroid size	
In school children 6-12 years of age: Proportion with enlarged thyroid, by palpation or ultrasound	< 5%
4. Neonatal TSH	
Proportion with levels > 5mU/1 whole blood	< 3%

Source: *Indicators for assessing Iodine Deficiency Disorders and their control through salt iodization. Report of a Joint WHO/UNICEF/ ICCIDD Consultation, 3-5 November 1992, WHO, Geneva. Document WHO/NUT/ 94.6.*

specified under the provisions of item A.15.01 of Appendix B of the PFA Rules, 1954. The ban is applicable only to the salt meant for human consumption and not applicable to the salt for animal consumption. Strict ban enforcement is as important and essential as the ban notification. The PFA Act is implemented by Food and Drugs Administration of the concerned states. The food inspectors are empowered to draw samples in accordance with the procedure laid down in the rules and get them analysed in the designated laboratories. If a salt sample fails to meet the specifications, either in terms of iodine content or other characteristics, the food inspectors are empowered to launch prosecutions against the manufacturer / producer / retailer / seller of such substandard iodized salt. A successful prosecution could mean imprisonment which shall not be less than six months, and may be extended up to three years and a fine of not less than Rs 1000.

The full ban notification has been in force in 21 states and 5 union territories (table 4.1). Gujarat and Tamil Nadu have recently issued full ban notification with effect from 1 January 1995. Kerala, Goa, and Pondicherry have not issued any ban notification whereas Karnataka, Maharashtra and Orissa have issued partial ban notification.

Agenda for the future

Necessary efforts to ban uniodized salt in all states and union territories should be made. At present, the ban is only on salt meant for human consumption and not on salt meant for animal consumption. This implies that both types of salt are available in the market and the uniodized salt being cheaper is more likely to be bought by a not so knowledgeable consumer. Therefore, it is necessary that the ban is extended to all salts. It is also important to note that the IDD affects livestock as well and causes abortions, still births, decreased yield of meat, milk, egg, wool, etc. and a decreased capacity to work.

Capacity Building/Training

Current status

All India Institute of Medical Sciences, National Institute of Communicable Disease at New Delhi and All India Institute of Hygiene and Public Health at Calcutta have been identified as nodal training centres for both medical and non-medical personnel, monitoring of iodine contant of salt and urinary iodine.

Agenda for the future

Training about IDD should also be a part of the general training of the health workers at all state Health and Family Welfare Training Centres. Sufficient emphasis on IDD should also be a part of the medical undergraduate curriculum.

Issues Related to Salt Iodisation

Current status

The total annual production of salt in India is 13.5 million tonnes, out of which 5.2 million are used for human and animal consumption and the rest for industrial purposes. The three major salt producing states are Gujarat (68%), Tamil Nadu (15%) and Rajasthan (12%). There are, at present, about 6450 medium and large scale producers who account for 92 per cent of the total salt produced in the country, the rest 8 per cent being accounted for by about 3000 small-scale producers.

Of the total installed capacity of 6.6 million tonnes for salt iodization, only about 3 million tonnes are being produced. The salt requirement for the country is around 5 million tonnes and the rest can be exported. The production and demand for iodized salt can be raised substantially once all the states notify bans of uniodized salt and strictly enforce it.

To facilitate easy and speedier distribution of salt, the Salt Commissioner's Office and the Ministry of Railways have developed a railway zonal scheme linking production centres and sources with the nearest distribution points.

Agenda for the future

A good proportion of the salt is produced by medium and small-scale producers. They find it difficult and uneconomical to iodize their salt. Designing and production of hand-operated iodization plants at low cost (Rs 15,000 per plant) with a capacity to iodize about 100 kg salt/hour is being developed by an Indian firm. The transportation by road is increasingly being used and a need for proper channelization of routes is needed.

To prevent iodine losses during storage and transport, there is a need for safe and protective packing using HDPE or LDPE bags. In places, where buying is annual/semi-annual, one should advocate large LDPE bags rather than small packets. With the increasing concern of environmentalists about the use of plastic bags, there is need for research to look for a biodegradable yet protective material that meets the packaging requirements of consumers.

Information, Education and Communication

Current status

Television and radio spots addressing different issues in IDD have been agreed upon and are being broadcast. These materials have been prepared after extensive Knowledge Attitude Practice (KAP) surveys in the affected areas. These messages are primarily aimed at the general public and are meant to create a demand for iodized salt.

As a policy decision in 1989, the Government of India has decided to promote the national logo of 'smiling sun' for an easy identification of iodized salt by the public. If properly enforced, this not only ensures a quality product but also gives the public value for their money.

Agenda for the future

There have been repeated concerns about the toxicity of iodized salt and the need for universal iodization. Issues related to the safety of universal salt iodization have been carefully examined by the WHO which recommends that daily intakes of up to 1mg (1000 μg) appear to be entirely safe. At the current recommended levels in India, the daily intake is likely to be in the range of 150 μg to 300 μg. The benefits to be derived from universal salt iodization by the affected, and the absence of significant adverse effects among others in the same area who are not iodine deficient, far outweigh any risk of excess intake for a small minority.

IEC efforts aimed at doctors, nutritionists and decision-makers are necessary to remove the misconceptions regarding iodine toxicity. Various professional bodies of doctors, pediatricians, nutritionists, endocrinologists like the Indian Medical Association should include this in their agenda.

Monitoring

Current status

Success of the National IDD Control Programme depends primarily on the iodine level in salt at the retailer level, and finally at the consumer level. For an effective salt iodization programme, monitoring the levels of iodine at various stages and under various circumstances in one of the most important factors.

Every iodized salt manufacturer has to maintain a process control laboratory to ensure a proper level of iodization. A second check is carried out by the Salt Department before the dispatch of salt by rail or road.

Quality control at the retail level is done by the State Governments through the Health Directorates. The Food Inspectors of the concerned State Governments maintain the quality of iodized salt. In order to strengthen the monitoring system, the Government of India with UNICEF assistance has finalized a project for intensive IDD activities in four endemic states Assam, Madhya Pradesh, Himachal Pradesh and Uttar Pradesh.

Though the government has a system for monitoring, it is necessary to supplement it by external agencies. Non-governmental organizations (NGOs) have an important role to play in this and its success has been proven by a recent study undertaken in UP. The study was done in sixteen districts around Dehradun, Gorakhpur and Varanasi. Three NGOs were selected who further identified more peripheral NGOs who collected salt from the households and retailers. It revealed that out of 10,153 salt samples collected, 60.9 per cent of the salt samples had iodine levels below the recommended 15 ppm. The iodine level was especially low in Gorakhpur and Varanasi zones where 72 per cent to 73 per cent of samples had iodine levels below 15 ppm the compared to 50 per cent in Dehradun. A total of 5.3 per cent salt samples had no iodine.

Agenda for the future

Though the above study proved the feasibility of using NGOs for this purpose, more NGOs in different parts of the country have to take up this challenge. With the availability of spot testing kits, monitoring can be done in schools and even the general public can be asked to participate in this activity. Criteria for monitoring progress towards the elimination of IDD as a public health problem is given in table 4.4.

Community Participation

Current status

Modest efforts have been made by some NGOs in this field using their existing network. However, much needs to be done. With the promulgation of the

Panchayati Raj Act, community participation can actually be demonstrated in practice. Madhya Pradesh, which is the first state of the country to implement this Act, has already demonstrated the immense potentiality of this scheme. In conjunction with the high level of political commitment, Madhya Pradesh has forged ahead with an action programme to eliminate IDD from the state by 1997.

Agenda for the future

In the continuing process of the empowerment of people to take their own decisions, IDD should feature as an important issue. The impact of iodine deficiency on intelligence and educability of children is an important impediment for the development of a community. It is also serious handicap for our efforts to educate all by the end of this century. It is absolutely essential that all the impediments are removed in its path. If necessary, the consumer forums and NGOs should be encouraged to file public interest litigation on behalf of cretins and the children of India. It is indeed paradoxical that at present some people/organizations are questioning universal salt iodization.

Building a Partnership

There are a number of contributors to the successful elimination of IDD. Each will need to understand the potential of the others. These stakeholders include the government, the food processing industry, nutritionists, educationists, agriculturists veterinaries and animal husbandry workers, departments, communications professionals and people working with the mentally disabled. Commitment to, and ownership of the solution to the IDD problem by this wider network of stakeholders will be essential to ensure the elimination of IDD by the year 2000.

This agenda of involving and building alliances with a large and diverse range of interested groups is ambitious but not impossible. The first step in this direction was taken in April 1995 when delegates of six South Asian countries—Bangladesh, Bhutan, India, Nepal, Pakistan and Sri Lanka came together to decide the strategy for the subcontinent.

Sustainability

As said earlier, unlike communicable diseases, eliminating IDD is not a one time effort. The main effort is to sustain the achievements. Many countries have suffered setbacks in their programme of IDD elimination, when success induced a sense of complacency.

The iodization of salt, while undoubtedly a simple technique for an immeasurably important social benefit, is, nevertheless, only the first step in the goal of elimination of IDD. Many other steps will be needed before IDD can be prevented from ever making a comeback. The strategy for the management of sustained IDD elimination should, therefore, focus on continuous political and financial commitment, clear communication strategy, persistent quality control, regular monitoring and a responsive administration.

References

Bleichrodt, N., M.P.,Born, 1994. 'A meta analysis of research on iodine and its relationship to cognitive development'. In: Stanbury,J., (ed). *Iodine deficiency and brain damage : Report of Franklin Symposium.* (in press).

'Citizens' report on iodine content in salt at consumer level in Uttar Pradesh'. Published by Centre for Community Medicine, All India Institute of Medical Sciences, New Delhi.

Department of Health and Family Welfare, Government of Madhya Pradesh, 1995. 'Rajiv Gandhi Mission for elimination of Iodine Deficiency Disorders: Mission Document', Bhopal.

'ICMR task force study : Epidemiological survey of endemic goitre and endemic cretinism'. Indian Council of Medical Research, New Delhi; 1989.

'International Council for Control of Iodine Deficiency Disorders (SEARO). Proceedings of Technical review meeting on iodine deficiency disorders in livestock populations'. G.B. Pant University for Agriculture and Technology, Pantnagar, Dist. Nainital, Uttar Pradesh, India; 27-28th January 1995. (in press)

Kochupillai, N., M.M., Godbole, C.S., Pandav, 1984, et al. 'Neonatal thyroid status in iodine deficient environments of the sub-Himalayan region'. *Indian J Medical Research.* 80: 293-8.

Ministry of Industry, 1994. 'Universalization of access to iodised salt—A mid-decade goal'. *The Salt Department, Ministry of Industry.*

Ministry of Human Resource Development. National Plan of Action on Nutrition. Food and Nutrition Board, Department of Women and Child Development, Ministry of Human Resource Development, Government of India, New Delhi. 1995.

Ministry of Industry. 'Banning sale of edible non-iodized salt – An urgent measure'. Produced by The Salt Department, Ministry of Industry with support from UNICEF, 1995.

Pandav, C.S., H., Vishwanathan, D.P., Haxton. 'Sustaining elimination of iodine deficiency disorders'. *National Med J India (editorial)*, 1995; 8 (2) : 49-50.

Sankar, R., T., Pulger, Bimal Rm et al. 'Iodine Deficiency Disorders in school children of Sikkim'. *Indian Pediatr* 1994; 61 : 407-14.

'UNICEF Progress towards universal salt iodization : An update on the current status of universal salt iodization in countries where UNICEF has programmes'. Nutrition section, UNICEF, New York, December 1994.

World Health Organization, 1993. WHO/UNICEF/ICCIDD Global prevalence of iodine deficiency Disorders. 'Micronutrient Deficiency Information system'. MDIS Working Paper # 1.

WHO/UNICEF/ICCIDD 'Indicators for assessing Iodine Deficiency Disorders and their control through salt iodization'. WHO/NUT/94.6.

'World Health Organization : Iodine and Health : Eliminating iodine deficiency disorders safely through salt iodization-A statement' by the WHO, WHO/NUT/94.4; Geneva, August 1994.

PART II

Livestock and the Indian Economy

5

Role of Livestock in the Indian Economy

S.S. Khirwar and K. Pradhan

Abstract

Livestock has complementary, supplementary, and sustainable relationships with crops under the mixed farming systems that are prevalent in the country. The diversity of livestock species has been increasing, especially since the Fourth Five Year Plan (1969-74) when the breeding policy was changed to crossbreeding. The majority of livestock, however, is owned by landless, marginal, and small scale farmers who are categorized as the poorest of the poor in the society. The livestock sector provides valuable animal protein for the human diet in the form of milk, meat, eggs, and chicken. It also provides draught power for agricultural operations and rural transportation, dung for maintaining soil fertility and energy for cooking in rural homes. Wool, hair, hides, skins and bones are some of the by-products from animal enterprises which are used as raw materials for garments, carpets, shoes, and the mineral industry, all of which help in earning valuable foreign exchange. Livestock generate employment for the rural population, particularly for women. It serves as a source of ready cash and is least affected by many calamities. The value of output from livestock has increased five-fold during the period from 1980 to 1990, but investment in this sector has declined from 0.81 to 0.58 per cent of the total plan expenditure. This sector offers opportunities for the improvement of farming technology in the production of better livestock suitable to different agro-climatic zones of the country, improvement of animal drawn vehicles and implements, post-harvest technology for livestock products, improvement of information and communication systems for the transfer of technology in the animal sciences, and strong programmes of human resource development.

Introduction – Agriculture and Land

Agriculture continues to be the most dominant sector of the Indian economy, even though, its contribution to the gross domestic product (GDP) has declined to 32 per cent. It provides employment to 196.77 million rural people in the country. Livestock rearing is an important constituent of the agriculture sector in the country. Land, which is the basic resource in farming, is distributed very inequitably; its fragmentation is increasing with the passage of time. For instance, an analysis of the national sample survey data on land holdings between 1971-2 and 1981-2 shows that the number of marginal holdings rapidly increased with a sharp decline in the average size (Nair and Dhas, 1989). The majority (55%) of the holdings were below one hectare and approximately 19 per cent of the holdings were between one to two hectares. These holdings accounted for only 10.7 - 12.8 per cent of the

Fig. 5.1. Sheep with goitre in a region of iodine deficiency. (Photo E. Cabezas - Reproduced with permission from — Salt Iodization for the Elimination of Iodine Deficiency by M.G. Venkatesh Mannar and John T. Dunn a ICCIDD/MI/UNICEF/WHO Joint publication, 1995).

total cropped area. In addition, about 30 per cent of the rural households were landless. This situation describes the basic problem in taking up intensified agriculture on a larger scale in India. The distribution of livestock on different holdings indicates that the number of bovines (cows, buffaloes) as well as ovines, increased in 1986-7, as compared to 1981-2 in all holdings (irrespective of size) indicating that the importance of livestock husbandry as a sub-sector of agriculture has increased over the years. Mixed farming systems that incorporate dairy animals are better suited for small-holding farmers as animal husbandry activity provides opportunities for utilizing surplus labour on the farm, in addition to generating a much needed regular cash supply. There is a ready market available for milk, meat and eggs which are always sought after by the urban population.

Livestock Wealth of India

The livestock resource of India in characterized by genetic diversity spanning resistance to tropical diseases, tolerance to very high and low temperatures and humidity, and good quality beef, veal, and milk production traits. Some of the finest draught cattle breeds and top ranking buffalo germ plasm is being exported from India to other countries for the development of their local livestock.

Cattle and buffalo have been integral parts of the crop-livestock mixed farming system of the Indian farm economy for thousands of years. Possessing nearly 271 million bovines, India ranks first in the world with 19 per cent of the world's bovine population and 51 per cent of the Asian bovine population (FAO, 1993). Almost all (97%) of the world's buffalo population is in the Indian sub-continent and India is home to 53.3 per cent of the world's buffalo population. The ratio of cattle to buffalo population approximates 100 to 11 globally, 100 to 34.6 in Asia, and 100 to 38.1 in India.

The total cattle population increased from 155.3 million in 1951 to 186.5 million in 1981, registering an average annual compound growth rate of 0.70 per cent. During the subsequent decade, however, the growth rate of the cattle population was very slow, with a combined growth rate from the year 1951 to 1990 of only 0.45 per cent (table 5.1).

The buffalo population increased from 43.4 million in 1951 to 66.1 million in 1981, registering an average annual compound growth rate of 1.54 per cent. The combined growth for the period from 1951 to 1990 is 1.49 per cent, indicating that the buffalo population is growing at a faster pace in comparison to cattle. This is mainly due to the fact that the buffalo has become the animal of choice in most of the milkshed areas of northern and western India. It is interesting to note that breedable cows are growing at the rate of 0.70 per cent in comparison to the total cattle population, registering an annual growth rate of 0.45 per cent. In the case of buffalo, the growth of the total population (including breedable buffaloes) is 1.49 per cent. This further strengthens the view that the buffalo is more popular than the 'sacred' cow in India for the production of milk.

The sheep population increased from 39 million in 1951 to 45 million in 1981, a growth rate of 0.25 per cent per annum, in spite of the large scale slaughter of these animals for mutton production. During the next

Table 5.1. Livestock and poultry population in India (in millions)

Species	Year					Compund Growth Rate (% per annum)
	1951	1981	1990	1991*	1992*	1951-90
Cattle	155.3	186.5	191.7	193.3	192.6	0.45
Buffalo	43.4	66.1	78.3	79.0	78.6	1.49
Pig	4.4	9.4	10.4	10.4	10.5	2.07
Sheep	39.0	45.0	43.7	44.2	44.4	0.25
Goat	47.1	82.0	110.0	116.8	117.0	2.18
Camels	0.6	1.0	1.4	1.5	1.5	2.04
Horses	—	0.9	0.9	1.0	1.0	---
Asses and Mules	—	1.1	1.6	1.6	1.6	---
Fowl**	73.5	160.0	350.0	380.0	410.0	4.16

*FAO estimates

** Chickens, ducks, turkeys

Source: *FAO Production Yearbook Vol. 46. Food and Agriculture Organization: Rome. 1993.*

decade, the number declined to 43.7 million. Even then, the country ranked fifth in the world in sheep population. Extensive crossbreeding with Rambouillet has led to superior strains, such as Kashmir Merino, Avivastra, and other fine wool hybrids developed at the Central Sheep and Wool Research Institute (CSWRI), Avikanagar, Rajasthan, as well as in Gujarat, and Tamil Nadu. Similarly, hybrids involving Dorset Horn and local sheep have shown promising results for meat production (Rao, 1992). This technology which has been evolved at different research stations, if transferred to the flock owners, could result in increased meat production for export.

The goat population, which stood at 47.2 million in 1951, increased to 67.5 million in 1972 and 82 million in 1981 (Table 1). The growth of the population was the highest amongst cattle, buffalo, and sheep. During the period from 1951 to 1980, the annual average compound growth rate was 2.38 per cent; however, for the combined period from 1951 to 1990, this rate was 2.18 per cent. This growth rate was maintained even when 35 per cent of the goats were slaughtered per annum. Goats can withstand very harsh environments and the country has some important milch goat breeds, such as Jamunapari, Beetal, and Sirohi.

Pigs are considered to be one of the most efficient converters of feed and have been most neglected. The population has increased from 44.3 lakh during 1951 to 104 lakh during 1990 showing a 237 per cent increase during 40 years. The compound growth rate works out to 2.07 per cent. Pork is considered to be the poor man's meat in India and offers opportunities of providing a necessary component of the human diet at a low price, especially when the prices of mutton and chevon are high, taking these commodities out of the reach of the common man.

Camel, horse, asses, and mules are essentially used as draught and pack animals and do not contribute to the food supply in India. Their numbers have also increased with the passage of time. The camel population has grown at a compound annual growth rate of 2.04 per cent per annum during the period from 1951 to 1990.

The poultry industry has shown remarkable progress, unparalleled by any other sector. The poultry population, which stood at 73.5 million during 1951, has increased to 410 million with an average annual growth rate of 4.16 per cent. This spectacular growth has been achieved with very heavy investment. Egg and broiler production have increased at similar rates. Even though poultry has to compete in its dietary needs with the human population, this increase in egg and broiler production is commendable and shows dynamism in the poultry industry.

The foregoing discussion has clearly brought out that the country has a huge livestock population which needs heavy investment to maintain production. This is essential for providing economic stability to the poor, cash-hungry farmers who possess small land holdings, particularly in dry/rainfed areas which constitute 70 per cent of the arable land of the country.

Livestock Ownership

Although ownership of livestock by different categories of farmers is not available on an all-India basis, the results of surveys conducted by National Dairy Development Board (Khanna, 1989), show that nearly 60 per cent of rural households own milch animals

Table 5.2. Land and milch animal holdings

Category	Landless	Marginal	Small	Medium	Large	All
Families	2.131	1.690	1.025	0.675	0.508	6.029
(millions)	(35)	(28)	(17)	(11)	(9)	(100)
Families owning milch	0.770	1.052	0.799	0.559	0.445	3.625
animals (millions)	(21)	(29)	(22)	(16)	(12)	(100)
Animals	1.390	2.091	1.932	1.651	1.877	8.941
(millions)	(16)	(23)	(22)	(18)	(21)	(100)
Animals* per owner	1.8	2.0	2.4	2.9	4.2	2.5
(ones)						

Figures in parentheses are percentages of the total
*Buffaloes and cattle: males and females of all age groups

Source: *Khanna, R.S. 1989.A historical perspective of dairy development in India. Indian Dairyman. 41:1-17.*.

(table 5.2). A total of 72 per cent of families owning milch animals were landless, small and marginal farmers, while a total of 61 per cent of the total cows and buffaloes in the area were held by landless, small and marginal farmers. Analysis of survey data (GOI, 1993) revealed that nearly 115 million bovines and nearly 50 million ovines were owned by farmers who had less than one hectare of land (tables 5.3 and 5.4). Although the number of bovines and ovines increased with the size of the holding, the highest number of livestock were held by farmers having smaller holdings. This trend, observed during the year 1981-2, continued untils 1986-7 when the number of livestock per holding increased tremendously within a span of five years. Small and marginal farmers keep animals mainly to supplement their income by utilising the extra time available after their main occupation (i.e., mainly crop husbandry in the countryside). These figures do not provide estimates of livestock ownership by landless families.

Various studies (NDRI, 1994) conducted in Uttar Pradesh, Maharashtra, Haryana, and Gujarat have indicated that most of the landless families in the rural areas are engaged in some livestock enterprises by which they are able to generate incremental income. Most of the data for these studies have been collected for cattle and buffalo only. Sheep, goat, and pig rearing, however, occupation of landless families which constitutes the poorest of the poor in even the agriculturally advanced States of the country. A study conducted in Bihar (Singh and Kumar, 1994) revealed that 78.8 per cent of goat owners were landless, marginal and small farmers.

During the last decade, poultry farming has become a commercial venture in different states of the country. Broiler and egg production have increased

Table 5.3. Distribution of livestock according to the size of holdings during 1981-2

Size of Holding	No.of Holdings (Thousands)	Area per Holdings (ha)	Livestock per Holding	
			Bovine	Ovine
Below 1.00 ha	46,728	0.43	1.57	0.94
1.00 - 1.99 ha	16,016	1.41	3.82	1.51
2.00 - 3.99 ha	11,966	2.72	4.73	1.73
4.00 - 9.99 ha	7510	5.96	6.38	2.04
10.00 & above	1871	16.20	9.36	3.38
All size classes	84,091	1.79	3.33	1.31

Source: *GOI Government of India Report of the Technical Committee of Direction for Improvement of Animal Husbandry and Dairying Statistics. Animal Husbandry Division, Department of Agriculture & Co-operation, Ministry of Agriculture, New Delhi, 1993.*

Table 5.4. Distribution of livestock according to the size of holdings during 1986-7

Size of Holding	No.of Holdings (Thousands)	Area per Holding (ha)	Livestock per Holding	
			Bovine	Ovine
Below 1.00 ha	50,291	0.40	2.33	1.05
1.00 - 1.99 ha	16,407	1.45	3.70	1.77
2.00 - 3.99 ha	12,373	2.76	5.47	1.94
4.00 - 9.99 ha	7427	5.90	7.29	2.36
10.00 & above	1830	14.03	9.16	4.85
All size classes	88,958	1.66	3.70	1.49

Source: *Same as table 5.3.*

manifold due to the trasformation of poultry farming in to industry. Several big concerns have arisen, particularly in the States of Haryana, Andhra Pradesh, Tamil Nadu and Karnataka where poultry farming used to be a small backyard activity but has now been transformed into an industrial venture.

Investment in the Animal Husbandry Sector

Keeping in view the role played by the animal husbandry sector in providing valuable protective foods (particularly milk, meat, eggs, etc.) to the increasing human population, the government has increased the fund allocations to this sector in the successive five year plans (table 5.5). During the First Plan, agriculture had 14.8 per cent of the total plan expenditure, while only 0.81 per cent (i.e., Rs. 160 million: 5.5 per cent of the expenditure on agriculture) was spent on animal husbandry and dairying. Although the share of agriculture in plan expenditure decreased to 12.5 per cent of the total plan expenditure, the total amount spent on agriculture was Rs 1,36,203 million during the Sixth Plan (1982-5). This amount was 45 times more than the expenditure in the First Five Year Plan (1951-6). At the same time, the expenditure on animal husbandry increased by 50 times from Rs 160 million during 1951-6 to Rs 8025 million in the Sixth Plan. During the Seventh Plan (1985-90), allocation for agriculture has virtually remained static at 12.7 per cent of the total allocations. The expenditure on animal husbandry and dairying has been reduced to 0.58 per cent in comparison to 0.81 per cent of the total plan

Table 5.5. Investment in animal husbandry sector during various plans (Rs in million)

Plan Period	Total Plan	Agriculture and Allied Activities*	Expenditure on Animal Husbandry and Dairying*
First Plan (1951-56)	19,600	2,900 (14.8)	160.0 (0.81)
Second Plan (1956-61)	46,720	5,490 (11.7)	334.7 (0.72)
Third Plan (1961-66)	85,765	10,890 (12.6)	770.0 (0.90)
Fourth Plan (1969-74)	157,788	23,204 (14.7)	1,542.6 (0.99)
Fifth Plan (1974-78)	394,262	48,665 (12.3)	2,324.6 (0.66)
Sixth Plan (1980-85)	1,092,917	136,203 (12.5)	8,025.1 (0.80)
Seventh Plan (1985-1990)	2,202,163	279,611 (12.7)	12,805.0 (0.58)
Total	4,187,163	533,031	28,646.7

* Figures in parentheses denote percentage of the total plan expenditure

Source: *Same as table 5.3.*

Table 5.6. Production of major livestock products

Year	Milk (Million Tonnes)	Eggs (Millions)	Wool (Million kg)
1950-51	17.0	1.832	27.5
1960-61	20.0	2.881	28.7
1973-74	23.2	7.775	30.1
1980-81	31.6	10.060	32.0
1989-98	51.4	20.204	41.7
1990-91*	54.9	21.343	42.7
1991-92**	56.3	21.723	41.6
1992-93*	58.6	23.067	43.3
1993-94***	61.0	24.809	45.6
1996-97***	70.8	30.001	50.1

*Provisional
**Anticipated Achievement
***Target

Source: *Same as table 5.3.*

expenditure during the First Plan. The anticipated Eighth Plan expenditure has been put at Rs 13,367 million for the animal husbandry sector, which is almost equal to the total expenditure for the entire 40 year period from 1951 to 1990. When the expenditure on animal husbandry and dairying is expressed as a percentage of the expenditure on agriculture and allied activities, it was observed that the percentage decreased, starting from the Fifth Plan. During the Seventh Plan, it was 5.8 per cent while in the Eighth Plan, it was only 4.6 per cent. This decline in expenditure in the animal husbandry sector is contrary to the growth in this sector.

An estimated Rs 10,500 crores is the capital investment in the poultry industry which includes the food industry. The value of eggs and poultry meat produced annually is Rs 20,500 crores. Per capita egg availability has come up from 5 in 1951 to 28 in 1990. All this has been achieved due to the transformation of traditional poultry farming into a dynamic agricultural business. The growth achieved has surpassed all sectors, including industrial sectors.

Production of Major Livestock Products

The livestock sector provides milk, meat, eggs, wool, organic manure and much needed draught power for the small holdings in rural areas. The major share of the production comes from milk, eggs, meat (including meat by-products), and wool. The production of major livestock products, namely milk, eggs and wool, is presented in table 5.6 with their growth rates in table 5.7.

Table 5.7. Growth rates in the production of major livestock products (% per annum)

Period	Milk	Egg	Wool
1950-51 to 1960-61	1.64	4.63	0.43
1960-61 to 1973-74	1.15	7.91	0.37
1973-74 to 1980-81	4.51	3.79	0.88
1980-81 to 1990-91	5.68	7.81	2.93

Source: *Same as table 5.3.*

Table 5.8. Per capita supply of animal products in India

Year	Milk (g/day)	Meat (kg/year)	Chicken (g/year)	Egg (no./year)
1980	125	1.77	190	18.9
1985	153	1.82	241	21.6
1988	160	2.14	354	25.6
1989	164	2.28	403	28.0
1990	166	2.37	444	30.0

Source: *Same as table 5.1.*

Milk Production

Milk production, which was 17 million tonnes during the year 1950-1, doubled in a span of 30 years. This production figure rose to 51.4 million tonnes during 1989-0, which is three times more than that for the year 1950-1. The cumulative growth rate for the first 10 year period was 1.64 per cent, which decreased to 1.15 per cent during 1961-74. The growth rate leaped to a level of 4.5 per cent during the year 1974-80, which further increased to 5.68 per cent during the period from 1980-1 to 1990-1. The target for 1993-4 has been fixed at 61 million tonnes while the same for 1996-7 has been placed as 70.8 million tonnes. The growth in milk production after the year 1973-4 was mainly due to the launching of animal husbandry and dairy development programmes with particular reference to the 'Operation Flood' programme and an intensified cross-breeding programme taken up by different states during this period. When milk production is staggered over the different regions of the country and Union Territories, it was found that during 1970-1, 50 per cent of the total milk produced came from the northern region followed by the western (20.2%), southern (16.9%) and eastern region (12.1%) (Patel, 1993). These figures changed to 43.7 per cent, 21.7 per cent, 20.1 per cent, and 13.7 per cent, respectively during the year 1989-90 which indicates that the growth rate during the period from 1970-1 to 1989-90 was highest in the southern region (5.92%) followed by eastern (5.45%), western (5.02%) and northern region (4.04%).

During the period from 1981 to 1990, India's cow milk production increased at a combined growth rate of 7.29 per cent per annum as compared to 1.02 per cent in the world and 3.94 per cent in Asia. Buffalo milk production showed a national increase of 3.72 per cent as compared to 3.93 per cent in the world and 4.09 per cent in Asia. Although the share of goat milk in total milk production in India is meagre (2.91%), its production showed an annual growth rate of 5.81per cent as compared to 2.35 per cent in the world and 2.22 per cent in Asia (Patel, 1993). India ranks sixth in the world in cow milk production and is a leader in buffalo and goat milk production.

Egg Production

Egg production, which was 1832 million during the year 1950-1, increased to 2881 million in the year 1960-1 with a corresponding compound annual growth rate of 4.63 per cent. This growth rate increased to 7.91 per cent between the period from 1960-1 to 1973-4. During the period from 1973-4 to 1980-1, the growth of egg production was slower but production of 10,060 million eggs was achieved. During the last decade (i.e., between 1980-81 to 1990-91), the growth of egg production was 7.81 per cent per annum with a corresponding egg production of 21,343 million during 1990-1. The egg production target for the year 1993 and 1996 has been fixed at approximately 24,800 and 30,001 million eggs, respectively. When the data of egg production is considered on the basis of production in different states, it is observed that Andhra Pradesh produced the largest number of eggs (3700 million) followed by Tamil Nadu (2704 million), West Bengal (2327 million), Maharashtra (2145 million), and Punjab (1900 million). When per capita availability of eggs is worked out for the year 1989-90, Andaman and Nicobar Islands tops the list with an average availability of 123 eggs per annum followed by Goa (78), Punjab (74), Chandigarh (60), Andhra Pradesh (55), Kerala (51), and Tamil Nadu (42).

Wool Production

Wool production has registered very slow growth as compared to other livestock products. During the year 1950-1, wool production stood at 27.5 million kg which increased to 28.7 million kg during 1960-1, registering a compound growth rate of only 0.43 per cent per annum. During 1973-4, production was 30.1 million kg, while in the year 1980-1, it increased to 32 million kg. During the next decade, the growth rate of wool production was recorded as 2.93 per cent with 42.7 million kg wool. A target for wool production for the year 1993-4 and 1996-7 has been fixed at 45.6 and 50.1 million kg, respectively.

Meat Production

Meat production, which stood at 7.64 lakh tonnes during 1971, increased to 9.10 lakh tonnes during the year 1981. A spurt in meat production was recorded in the year 1989 when the production increased from 1.75 million to 3.10 million tonnes. This increase was mainly due to a spurt in beef and veal as well as a 400 per cent increase in buffalo meat production. During 1990-1, the production of meat from different sources stood at 3.28 million tonnes. The share of beef and buffalo meat was 56 per cent, followed by sheep and goat meat (18%), poultry (11.5%) and pig (11%). The figures for duck, turkey, quail, etc. were not available; however, altogether they constitute nearly 2.5 - 3 per cent of the total meat production in the country. Annual meat production has increased by 6.5 per cent during the period from 1980 to 1990. Meat production showed a phenomenal increase far exceeding the growth rate of other food material. With the increase in production, the availability of meat per person per year has increased from 1.7 kg during the year 1980 to 2.37 kg during the year 1990. The availability of chicken has increased from 190 grams per person per year during 1980 to 444 grams during the year 1990 (table 5.8). The production of other livestock by-products has increased with the increase in beef, buffalo, meat, mutton, chevon, and chicken production (FAO, 1993).

Contribution of Livestock to the Indian Economy

The value of the total output originating from livestock and agriculture enterprises during 1989-90 was estimated to be Rs 75,688 crores based on 1980-1 prices and the value of livestock products (i.e., milk, meat, eggs, wool, dung, hides, skins, etc.), was of the order of Rs 15,870 crores. This accounted for about 21 per cent of the total income which originated in the agriculture sector. This figure has increased to Rs 17,690 crores during the year 1990-1. On the basis of current prices, the value of livestock products was Rs 32,165 crores during the year 1989-90. The contribution of different livestock products in the total value of output from livestock is presented in table 5.9. It can be observed that milk from different species of livestock contributes nearly 75 per cent of the total value, followed

Table 5.9. Value of output from livestock in India (Rs. in crores at 1980-1 prices)

Item	Year				
	1980-81	1985-86	1988-89	1989-90	1990-91
Milk & milk products	6,884	9,530	10,517	11,188	11,711
Meat and meat products	1,570	2,110	2,412	2,519	2,632
Eggs	351	548	639	685	856
Wool & Hair	49	63	66	67	67
Dung	1,352	1,434	1,578	1,603	1,616
Silkwormm, cocoon, & honey	136	175	208	257	270

Source: *Same as table 5.3.*

by meat, poultry, and eggs. The value of dung, which provides valuable organic manure and saves the land from the vagaries of degradation, has been worked out to be nearly 10 per cent of the total output from livestock products. In addition to fulfilling the requirement of domestic industries, raw leather, meat, meat products, and finished leather goods are also exported to other countries. The value of such exports was Rs 27,680 crores during the year 1989-90, while this export increased to Rs 32,553 crores during 1990-1. This clearly indicates the significant contribution made by the livestock sector to the gross domestic product (GDP). The contribution of agriculture in GDP has been decreasing constantly due to increased industrial growth. The contribution of livestock has, however, remained static at about 20-22 per cent over the years. The share of the agriculture sector in the total GDP was 50.5 per cent in 1950-1. This share declined to 32.6 per cent by 1988-9 mainly due to a spurt in industrial growth in the country. The contribution of GDP from agriculture has gone down from 46.2 per cent to 37 per cent from 1980-1 to 1987-8 while the contribution from the livestock sector has increased from 22.89 per cent to 34.56 per cent. The annual market value of Indian livestock is around Rs 50,000 crores which is roughly equal to the food-grain produced annually (Khanna, 1989). The wool sector, in addition to meeting indigenous demands, exports annually Rs 500 crores worth of finished woollen products. Apart from this, hand knitted carpets and hosiery are also exported. About 29 million kg of fine wool are imported annually for re-export after value addition in the country.

Draught Power

Draught animals play a vital economic and social role in the Indian economy and in the lives of millions of farmers. They are woven into our social life in such a way that it is difficult to separate their role and assign economic value to them. They have become a part and parcel of our farming system. The population of draught animals is nearly 84 million. Assigning a value of 0.5 horse power for each draught animal gives a value of 42 million horse power energy, which is sufficient to cultivate nearly 125 million hectares and haul about 50 billion tonnes of load in nearly 15 million animal drawn vehicles. Working animals save nearly six million tonnes of petroleum fuel per year. This entails a saving of Rs 3000 crores in foreign exchange (Ramaswami, 1993). Major factors that will determine whether draught animal power will continue to be used in the country are considerations that: (i) bullocks are by-products of the milk production industry, (ii) poor farmers cannot invest the heavy capital for mechanization and so stick to traditional methods of cultivation, (iii) draught animals require less capital for maintenance, and (iv) draught animals utilize the by-products of agriculture which cannot otherwise be utilized. It has been estimated that in grain production enterprise, nearly 70 per cent of all products are not utilizable by man, although draught animals can be maintained on these by-products on the farm. Surveys recently conducted in Haryana (Chamola and Khirwar, 1993) indicate that tractors are more economical on large farms and can provide additional income to farmers by hiring them in the lean season so as to minimize post-harvest losses during farm operations. Draught animals, however, remain the mainstay for hauling. Cattle, which used to carry the load of agriculture in Haryana, have largely been replaced by male buffaloes. In short, the most important factors in the utilization of draught power in Haryana are the size of holdings, adoption of new technology, and the purchasing power of farmers. The share of animal energy, however, is constantly declining in the total energy matrix. Camels, horses, mules, donkeys, and elephants will continue to be used as draught and pack animals in areas where mechanical power cannot be applied, as in desert areas and hilly terrains where building roads would require heavy investment. By modernizing animal drawn vehicles, there is a tremendous scope for increasing employment and earnings for farmers during the lean season. Improved bullock carts and other animal drawn agriculture implements will help in increasing the use of draught power in agriculture.

Employment

The rural population increased from 290 million to 520 million between 1950-81. In 1987-8, the percentage of population living below the poverty level was 33.4 per cent in rural areas as compared to 20.1 per cent in urban areas (GOI, 1993). In 1972-3, 167.30 million rural people were engaged in agriculture, out of which 9.03 million were exclusively employed in the livestock sector. The number of persons employed in agriculture during the year 1977-8 increased to 183.40 million showing a combined overall annual growth rate of 1.96 per cent, while the number engaged in livestock increased to 10.60 million showing an overall growth rate of 7.41 per cent by 1987-8. The number of people employed increased to 16.65 million with a growth rate of 10.66 per cent during 1977-8 to 1983 and 18.31 per cent during 1983 to 1988 (GOI, 1993). Overall, the growth rate in employment in the agriculture sector from 1972-3 to 1987-8 was 1.1 per cent in comparison to 4.15 per cent in the livestock sector. The growth rate of employment has been rapid, especially after the year 1980.

Table 5.10. Value of output from agriculture and livestock sectors (Rs. in billions)

Year	Agriculture	Livestock
1980-1	589	106
1981-2	637	128
1982-3	669	144
1983-4	800	167
1984-5	850	196
1985-6	913	219
1986-7	975	246
1987-8	1089	280
1988-9	1329	322
1989-90	1461	377
1990-1	1727	437
1991-2	2018	516

Source : *Same as table 5.3.*

Conclusion

The productivity of several agricultural crops has reached a plateau. In order to break this stagnation, investments need to be very large. The productivity potentials in livestock and poultry, however, have not been fully exploited due to feeding and management constraints. Animal husbandry, a potential source of income to the rural poor, is now being recognized more than ever before by politicians, planners, professionals, scientists, and farmers. It is now a matter of identifying priorities for national development and upliftment of the rural poor who can actively participate in revolutionizing the animal husbandry sector. Scientists should help produce quality animals, poultry, feeds, and fodder, and maintain adequate health of livestock supported by improved management practices. Unless animals become more valuable to their owners, no economic success is envisaged.

It is essential that the benefits of research be passed on to the farmers. It is equally the responsibility of animal science research institutions to go into the field for the applications of research and let the farmers avail of its advantages. Gaps need to be sealed by reorienting the veterinary and animal science education and research to rural needs. Animal science research institutions must understand the needs of the farmers they propose to serve, so as to devise a system to train graduates that will reach farmers. The state agricultural universities and national institutions engaged in animal science research and education can make a significant contribution in providing these basic needs. It is also necessary to formulate appropriate policy and prepare plans on scientific lines for the development of the livestock and poultry sector, particularly at the level of the farmer.

The thrust areas envisaged for the improvement of production and productivity of livestock and poultry are: conservation and planned exploitation of germplasm resources; the improvement of farming technology for production of better livestock suitable to different agro-climatic situations of the country; improvement of post-harvest technology with an emphasis on storage, processing, and marketing of animal products; improvement with special reference to weaker sections of society; and advanced training to young scientists in the newly emerging fields of animal sciences.

In India, the 'green' and 'white' revolutions have contributed significantly to our economy and their advantages have tended to distribute evenly in the population as opposed to the polarization of wealth among a few due to industrial development. In a developing economy, a critical balance needs to be maintained between the growth of industrial and agricultural sectors.

References

Chamola, S.D. and S.S., Khirwar. 1993. 'Role of draught animal power in rural development'. In: Jain, C.M. and T., Cangan, (ed.) *Forty Years of Rural Development in India.* Printwell: Jaipur. pp. 84-99.

FAO Production Yearbook 1993. Vol. 46. 'Food and Agriculture Organization: Rome'.

GOI 1993. Government of India Report of the Technical Committee of Direction for Improvement of Animal Husbandry and Dairying Statistics. Animal Husbandry Division, Department of Agriculture & Co-operation, Ministry of Agriculture, New Delhi.

Khanna, R.S. 1989.'A historical perspective of dairy development in India'. *Indian Dairyman*. 41:1-17.

N.D.R.I. 1994. 'Proceedings of Seminar on Livestock for Sustainable Rural Employment and Income Generation'. Division of Dairy Economics, Statistics and Management, National Dairy Research Institute, Karnal.

Nair, K.N. and A.C., Dhas. 1989. 'Cattle breeding technology and draught power availability in India agriculture : An unresolved contradiction'. In: *ISAE Livestock Economy of India*. Oxford & IBH: New Delhi & IBH. pp. 13-44.

Patel, R.K. 1993. 'Present status and promise of dairying in India'. *Indian J. Agr. Econ*. 48: 1-33.

Ramaswami, N.S. 'Livestock and environment'. Proc. Indo German Conference on Impact of Modern Agriculture on Environment. Dec. 1-3, 1993, CCS, HAU, Hisar.

Rao, C.K. 1992. 'A case for a national policy on veterinary and animal sciences'. *Indian Vet. J*. 69: 7-25.

Singh, S. and S., Kumar. 'Goat enterprise for regular income and employment in Tribal areas of Bihar'. In: *Proceedings Seminar on Livestock for Sustainable Rural Employment and Income Generation*. Division of Dairy Economics, Stat. & Management. NDRI: Karnal. 1994, A-47

PART III

Iodine Deficiency Disorders in Livestock: Overview

6

Global Occurrence of Iodine Deficiency Disorders in Livestock

V. Kapoor and A. R. Rao

Abstract

Iodine is an essential element in the production of thyroxine, a hormone necessary for regulating growth and other metabolic functions. It is well known that production is seriously impaired in animals reared on iodine deficient rations. Poor productivity, reproductive disturbances, irregular estrus, and infertility have been reported in different species. Severe iodine deficiency in adult animals is known to cause weakness, hairlessness, and death. Milk production is positively correlated with iodine intake for many animals. Cereals are a poor source of iodine as compared to green fodders and their silages. The presence of goitrogens in feeds has been shown to inhibit the utilization of iodine leading to goitre even under adequate iodine intake. Thiocyanate, perchlorate, and certain elements (i.e. selenium and fluorine) in feeds or water known to cause goitre in animals. Thyroid size, iodine levels in urine, milk, serum precipitable iodine, protein-bound iodine, as well as the levels of thyroxine, and triiodothyronine are all useful indices of iodine status.

The potential for iodine toxicity is also discussed. This literature review considers baseline data from three reviews (Miller, 1975; Underwood, 1977, 1981) and reflects the last 20 year's worth of references on iodine from several journals on nutrition.

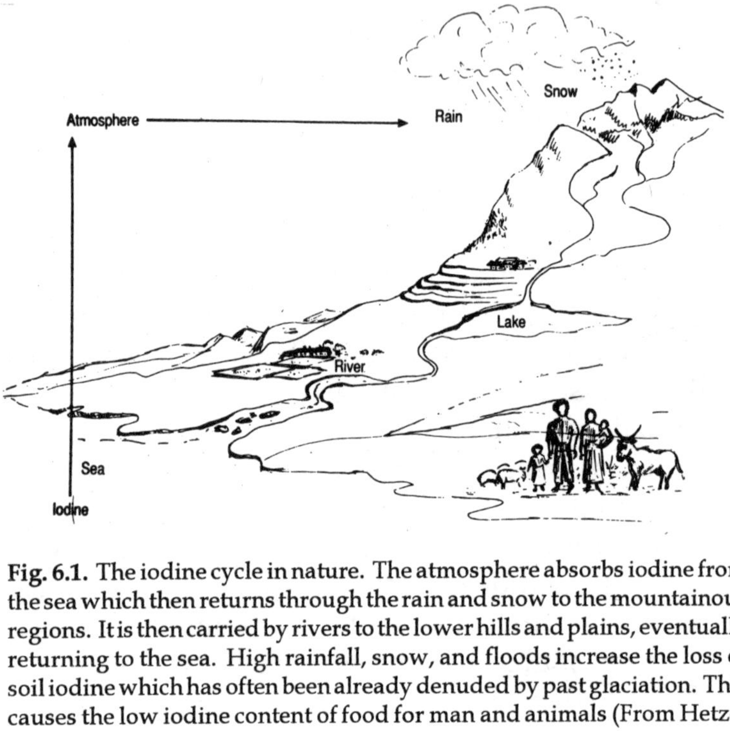

Fig. 6.1. The iodine cycle in nature. The atmosphere absorbs iodine from the sea which then returns through the rain and snow to the mountainous regions. It is then carried by rivers to the lower hills and plains, eventually returning to the sea. High rainfall, snow, and floods increase the loss of soil iodine which has often been already denuded by past glaciation. This causes the low iodine content of food for man and animals (From Hetzel 1989 with permission).

Introduction

In the animal body, iodine is found in both organic and inorganic forms with the highest organic concentration (70-80% of total) occurring in the thyroid gland and appreciable amounts in the ovaries, pituitary gland, and salivary glands. Inorganic iodine occurs in extremely low concentrations. It is reported that the normal range of plasma inorganic iodine in animals is 0.08 to 0.60 µg/dL. Values below 0.08 µg/dL are generally considered to reflect an iodine deficient status. Levels above 1 µg/dL are harmful to animals.

Extensive historical and scientific literature reveals that disorders of the thyroid gland are the most common endocrine disorders in humans. In livestock, nutritional iodine deficiency is the leading cause of thyroid gland disorders and its etiology begins with environmental iodine deficiency. The global prevalence of iodine deficiency is a geographical phenomenon mainly due to the low content of iodine in soil and water, precipitated by erosion and leaching. Glaciation, frequent flooding, distance from the sea, and annual rainfall are the principal causes of low iodine content in soils (Underwood, 1981). Iodine deficiency occurs in many species of animals living in iodine deficient areas (otherwise known as goitre endemic areas) of the world and manifests itself differently in different animals; productivity, however, is inevitably reduced in all animals with iodine deficiency. Alimentary tract motility has been shown to be reduced in hypothyroid monogastric animals [Rac *et al.* (1968); Falconer (1987); Underwood (1977)], as well as in ruminants (Miller *et al.* 1975). Iodine deficient calves may be born blind, hairless, weak, or even dead. These and other manifestations have a direct relationship with the severity of the deficiency.

Certain plant species containing variable amounts of goitrogens are known to cause goitre in animals by interfering with thyroid hormone synthesis. For example, prolonged consumption of the grass *Cynoden acthiopicus*, which is low in iodine and high in cyanogenic glycosides, is associated with a high prevalence of goitre in lambs (Rudert and Oliver, 1976). A survey conducted by Korber *et al*. (1989) in Germany revealed that about 10 per cent of cattle and 15 per cent of swine herds were iodine deficient mainly due to the presence of other goitrogens in their feeds, such as nitrates, thiocyanates, or glucosinolates.

Iodine Deficiency and Reproductive Disorders

The main function of the thyroid gland is to trap iodine and to synthesize, store, and release thyroid hormones. The manifestations of iodine deficiency, therefore, are those of a deficient supply of these hormones to the organism. Uptake of iodine from the blood is faster in the iodine deficient gland than in the iodine sufficient gland in its effort to maximize efficiency. Higher MIT/DIT and T_3/T_4 ratios in goats of goitre-endemic areas than in those from control areas can also be observed. All these responses reflect the compensatory mechanism of the thyroid gland to iodine deficiency (Karmarkar *et al*. 1974); however, biochemical and physiological functions are bound to deteriorate with time and severity.

Varying reproductive disturbances and irregular or suppressed estrus causing infertility have been reported in animals from goitre endemic areas.

Reproductive failure is the outstanding manifestation of iodine deficiency. Abortions, still births, and increased incidences of retained placenta are common. In rams, bulls, and stallions, decline in libido and deterioration in semen quality have been observed as the effects of iodine deficiency on male fertility.

Mavrin *et al.* (1976) demonstrated that iodine supplementation to animals in the USSR not only improved chances of conception but also shortened calving intervals and accelerated the onset of estrus. In a previous study, however, Timoshenko (1975), did not observe any increase in the fertility of cows in the USSR. Poor conception in cows has been reported in the forest and Amur region of the USSR where soils were deficient in iodine (Koval'skil, 1972). Supplementation of iodine at the rate of 0.5 mg/kg dry matter reduced the service period by about 50 per cent and improved the fertility of the cows by about 50 per cent. Feeding iodine at the rate of 6.0 -7.5 mg/cow daily resulted in a greater birth weight of calves. In Himachal Pradesh, India, supplementation of 10 ml Lugol's iodine in water daily for ten days to anestrous cows resulted in 62 per cent of the animals coming into estrus (Ryot *et al.* 1952). A relatively higher proportion (73%) of animals came into estrus when they received an iodine spray on the cervix in addition to Lugol's iodine supplementation. 64 per cent of the treated cows and 57 per cent of the treated buffaloes conceived.

It was reported in Australia that rams given 20 mg of KI twice weekly during the last three months of pregnancy gave birth to lambs having higher birth weights (Knights, 1979). Tasmanian workers reported a high mortality rate (36%) in Merino lambs with thyroid gland enlargement; a marked improvement in the birth weights and survival/growth rates was noticed when their mothers were fed with 20 mg of KI twice weekly during peak lactation (Knights *et al.* 1979).

Animal productivity has been reported to be impaired even in conditions of marginal iodine deficiency that does not cause clinical goitre. King (1976) reported a 36 per cent mortality in lambs from ewes showing minimal thyroid enlargement in Tasmania. Extra iodine administration to ewes in Queensland, Australia resulted in a marked improvement in birth weights and survival rates of Merino lambs (Knights *et al.* 1979). High rates of mortality and morbidity in new born calves in the low lands of Hungary and improvement in the cattle upon oral supplementation with missing minerals (including iodine) was observed by Bodal (1976).

Poultry appear to be able to withstand a considerable degree of iodine deficiency without marked loss of production or hatchability of eggs. On an iodine deficient diet for 35 weeks, hens did not show any effect on hatchability or embryo weight. After a period of two years, however, a decrease in hatchability, prolongation of hatching time, and retarded embryo development were observed (Underwood, 1981).

Iodine Deficiency and Animal Productivity

In reviewing the literature on iodine, Nyberg (1970) and Mason (1976) observed a linear relationship between the iodine content of milk and the prevalence of goitre. Iodine concentration of milk has been closely

Table 6.1. Ewe milk iodine vs. thyroid enlargement

Average Ewe Milk Iodine (μg/L)	Lambs with Enlarged Thyroid Glands (%)
23 - 64	23 - 69
71 - 98	12 - 22
98 - 111	0 - 4

Table 6.2. Average Iodine Concentration of milk and colostrum (5 Samples) of different Species

Iodine Concentration (μg/litre)

Species	Milk		Colostrum	
	Pantnagar	Pune	Pantnagar	Pune
Woman	34.20	78.02	97.03	152.09
Cow	32.59	72.65	89.09	143.41
Buffalo	42.51	78.54	112.81	174.34
Goat	50.06	81.33	154.16	228.19

Table 6.3. Iodine in cow milk with varying iodine intake

Iodine Intake (mg/cow/day)	Milk Iodine (μg)
1.6	28 ± 6
12.7	78 ± 18
20.0	267 ± 55

Table 6.4. Effect of supplemental iodine on milk iodine of cows

Iodine Intake (mg/day)	Milk Iodine (μg/L)	Range (μg/L)
68	694	608 - 870
6.8	81	51 - 124
no iodine supplementation	8	7 - 9

Source: *Hemken et al. 1972.*

linked with dietary iodine intake (Underwood, 1981; Iwarrson and Ekman, 1974; Cromwell *et al.* 1975; Berg *et al.* 1988) and so it has been suggested as an indicator of goitre status in a given area. For example, the level of iodine in animal milk in goitrous regions (20.6 μg/L) has been found to be lower than in non-goitrous regions (83.5 μg/L). Also, a decreasing level of iodine in ewe's milk was associated with an increasing prevalence of goitre in lambs, suggesting an inverse relationship (table 6.1).

Iodine content of milk and colostrum of woman, buffalo and goat from Pantnagar and Pune were analysed to study geographical variations. All the species at Pantnagar had lower levels of iodine in milk and colostrum than those at Pune (table 6.2). While Pune is about 100 km away from the sea, Pantnagar is in the foot hills of Himalayas, ten times farther from the sea.

Increase in milk iodine content due to increased dietary iodine has been reported (table 6.3). Miller *et al.* (1975) observed that approximately 10 per cent of dietary iodine goes into milk, however, this value may change with milk production. While reviewing the literature of two decades, Koiranen and Tancher (1976) observed that the iodine content of milk has increased appreciably due to the constant use of mineral supplements containing iodine. Hemken *et al.* (1972) reported an increase in milk iodine in cows from 8 µg/L to 694 µg/L with iodine supplementation (table 4). Addition of KI to the diet of lactating cows increased milk production by 10 to 18 per cent (Golubev and Sedov, 1984). Salakhutdinov (1985) estimated that iodine deficiency in the USSR resulted in the loss of 1,013,200 L of milk in 412 cows where the milk yield was 5.57 - 6.14 L per cow; normal milk yields were established by administering 18 mg of iodine per cow. S'edina (1987) reported that increasing the dietary iodine content of cows to the recommended level or 20 per cent above resulted in a 4.1 - 12.6 per cent increase in milk yield. Correction of iodine deficiency through supplementation with iodized salt has generally been shown to increase milk production from 4 -15 per cent. The minimum requirement of iodine in dairy animals has been reported to be around 0.6 ppm of feed dry matter (NRC, 1988). In water buffaloes, supplementation of iodine at the rate of 0.2 mg/kg body weight daily and twice weekly in their feeds for a period of seven weeks resulted in increased milk yield (Kobeisy and Shetaewi, 1992). Supplementation is especially effective in summer months (Budmaska, 1983).

Some reports suggest that iodine supplementations not only influence milk yield but also its composition. A decrease in volatile fatty acids in the rumen and an increase in milk fat is observed with

Table 6.5. Feed intake, milk production and body weight changes in cows fed two levels of supplemental iodine with a corn silage ration and soybean meal

Attributes	Iodine Supplementation (mg/day)		
	6.8	68	SE
Average days in milk prior to experimental period	37	39	--
Corn silage intake (kg/day)	20.5	21.3	0.61
Grain intake (kg/day)	11.3	10.8	0.37
Milk production (kg/day)			
Actual	24.8	23.5	--
Adjusted	24.3	24.1	0.79
Milk fat (%)			
Actual	3.65	3.44	--
Adjusted	3.64	3.45	0.10
Body weight gain (kg/day)	0.45	0.38	0.10

Source: *Hemken et al. 1972.*

iodine supplementation (Koval'kova *et al.* 1984; Kobeisy and Shetaewi, 1992; Budmaska *et al.* 1983; Kobeisy *et al.* 1992). Hemken *et al.* (1972), however, did not find any effect of iodine supplementation on milk fat as well as on feed intake or milk production of cows (table 6.5). Kobeisy *et al.* (1992) reported an increase in the percentage of protein, casein, lactose, calcium, and phosphorus in the milk of iodine-supplemented animals.

Increase in milk iodine may be due to factors other than the increased dietary intake of iodine. Dutch workers reported that iodine concentration in summer milk was low (9.7 µg/L) compared to winter milk (21.1 µg/L). Increased milk iodine due to teat dips has also been reported by various workers. A good amount of iodine is believed to be absorbed through the skin of the teats which is resecreted in the milk (Conrad and Hemken, 1978). Joerin and Bowering (1972) reported milk iodine content in the range of 10 - 700 µg/L when using an iodophor teat dip compared to 30 to 110 µg/L without a teat dip. Another study on farm animals showed that iodine content of milk ranged from 113 - 346 µg/L when teat dip were used compared to 13 - 23 µg/L without teat dips. Studies conducted by various workers (Dunsmore, 1976; Uvsi-Ranva *et al.* 1974), revealed an increased range of milk iodine (35 - 390 µg/L) using 5 g of iodine/L in the teat dip. Dunsmore *et al.* (1977) observed significant reduction in milk iodine when the teat dips had a low iodine concentration (i.e., 0.2 - 5.0 g/L).

Iodine concentration in milk decreases as lactation progresses (Groppel and Anke, 1983). Pyska (1994) observed that cow milk had at least 1.37 ± 0.1 µg of I/

Table 6.6. Iodine contents of soil, drinking water, feeds and fodders in different countries

Country	Soil (µg/kg DM)	Water (µg/L)	Fodder (mg/kg DM)	Feeds (µg/kg DM)	Reference
USSR	2.59 (1.5 - 4.09)	2.80 (2.30 - 3.43)	0.150 (0.02 - 0.58)	171 (9.8 - 660)	Salakhutdinov (1985)
USSR	0.099 - 0.434	---	0.043 - 2.0	---	Ogoleva *et al.* (1987)
New Zealand	1.5	---	< 0.1	---	Healy *et al.* (1972)
Ireland	2.92 - 30.15	---	0.08 - 0.42	---	McGrath and Fleming (1988)
Germany	—	1.3 (0.5 - 6.7)	---	---	Heinrich and Wenk (1989)

dL in the first three months of lactation as compared to 2.46 ± 0.18 μg I/dL at a later period, concurrent with a reduction in total milk yields.

The colostrum of cows has a higher level of iodine (200 to 350 μg/L) compared to the mature milk of the same animals (72-136 μg/L) (Miller, 1975). Iwarsson (1973), reported that the iodine content of cow milk fell during thc first few days immediately after parturition than at later stages of lactation. It is reported that the mammary gland of the sow has a greater ability to concentrate iodine than that of the cow.

Dietary Iodine Deficiency

Iodine content of soils, drinking water, feeds, and fodders in different countries are the main determinants of simple goitre in all classes of livestock (table 6.6). According to Golubev and Sedov (1984), soil and water samples in the USSR had iodine concentrations which were 44 per cent and 33 per cent below threshold values, respectively. Iodine concentration in drinking water is invariably much less in goitrous areas than in non-goitrous area as observed by various workers around the world (including India).

Svezhentsov *et al.* (1975) reported that many feeds from Moldavia were deficient in iodine. In south Germany, endemic goitre in farm animals has been reported to be due to the low iodine content of feeds (Heinrich *et al.* 1989). The range of iodine values in green fodders, like beet leaves and brassicas, is 73-196 μg/kg while in the legumes and graminae, it is 49-66 μg/kg. The average reported value of ensiled wilted red clover and sugar beet tops are 163 and 63 μg/kg dry matter, respectively. Iodine content of oat silage, maize silage, sugar beet top silage, grass silage, and red clover silage was 27, 38,74,76 and 91 per cent higher than in fresh material, respectively. Feeds prepared from maize silage, vetch and oat haylage, concentrates, sugar beet, vetch and oat hay, lucerne hay, and di-ammonium phosphate supplied iodine equal to 83.3 per cent of the recommended requirement.

Availability of herbage iodine is also affected by atmospheric temperature. Bedi (1991) reported that iodine content in berseem fodder grown at Izatnagar increased from 0.196 μg/kg in the first cut in December to 0.303 μg/kg of dry matter in the fifth cut in April. The probable reason suggested is that slow root nodule activity at low temperatures affects the mobilization of iodine and thus plant growth. Root nodule activity increases with rising temperature; however, a significant decrease is observed at temperatures above 30 °C.

Goitrogenic substances found in certain plants have a significant effect on aggravating iodine deficiency in various species of animals. Glucosinolate is the goitrogenic substance in plants which produces thiocyanate in the rumen of cows. The content of glucosinolate has been shown to vary in different varieties of kale. Reduced milk yield, neonatal mortality, and infertility or sterility due to irregular or suppressed estrus in lambs from ewes fed on goitrogenic kale was reported by Underwood (1981). Pregnant ewes fed a ration containing 20 per cent linseed meal produced goitrous lambs. Korber *et al.* (1983) observed thyroid disorders after 1.5 -2 per cent rapeseed meal (containing 67 μg of I/kg DM) had been incorporated in the ration of pigs.

Other goitrogenic factors can precipitate iodine deficiency in animals. The presence of dietary thiocyanates and perchlorates was shown to interfere with iodine uptake by the thyroid gland (Davilat and Skoknie, 1971). High fluoride intakes have also been contributing factors in the incidence of goitre in some parts of India and South Africa (Miller, 1975). Selenium is also a goitrogenic element. Goitrogenic effects, however, can be overridden by sufficient dietary iodine supplementation.

Iodine concentration in soil and herbage has not been found to be related to the distance from the sea (McGrath and Fleming, 1988). Groppel *et al.* (1989), however, reported that iodine content of feed and drinking water of a given area is influenced by distance from the sea. Nonetheless, iodine deficiency in soil, water, and feed is far more dependent on the topography, geological history, and environmental conditions of the area.

Requirements of Iodine

The minimum dietary iodine requirements of farm animals is difficult to express with accuracy as the requirements for growth may not be identical with those for reproduction and lactation, or for maintenance of the thyroid gland. Requirements of iodine for various species suggested by various workers are presented in table 6.7. Underwood (1981) had given the minimum iodine requirements of different classes of farm animals based on heat production, rather than energy production (table 6.8).

Iodine requirements are affected by diet, especially in the presence of goitrogens. Iodine should

Table 6.7. Iodine requirements of livestock

Animal	Iodine Requirement	Reference
Cattle, lactating (mg/kg DM of ingesta)	0.8	Newton and Clawson (1974)
Pigs, growing (ppm of DM of ingesta)	0.14	Shuman and Townsed (1963)
Pheasants and quail (ppm of DM of ingesta)	0.30	Scott *et al.* (1966)
Dairy cattle, lactating (ppm of DM of ingesta)	0.50	NRC, 1978
Dairy cattle, lactating (ppm of DM of ingesta)	0.60	NRC, 1988
Dairy cattle, non-lactating (ppm of DM of ingesta)	0.25	NRC, 1978
Dairy cattle, non-lactating (ppm of DM of ingesta)	0.25	NRC, 1988

*If the diet contains as much as 25% of a strongly goitrogenic feed on dry matter basis, iodine provided should be increased two times or more.

Table 6.8. Iodine requirements of farm animals

Animal	Body Weight (kg)	Heat Production (kcal/d)	Iodine Requirement (µg/day)
Poultry	2.3	225	5 - 9
Sheep	50	2,500	50 - 100
Pigs	68	4,000	80 - 160
Dairy cows	454	20,000	400 - 800

be at least double the requirements if the diet (on DM basis) contains 25 per cent of a strongly goitrogenic feed (NRC, 1978), such as soybean. Hemken *et al.* (1972) observed that 6.8 mg of supplemental iodine is adequate when a high soybean meal ration is fed with other foodstuffs containing about 0.9 mg of iodine/kg. Working with fresh forage diets, Barry *et al.* (1983) reported that iodine requirements of lambs can be met by a supplemented diet containing 180 to 270 μg of I/kg DM when kale (*Brassica oleracea*), ryegrass (*Lolium poerenne*), and clover (*Trifolium repens*) were the main constituents.

Salt is considered to be the most efficient vehicle for iodine supplementation to animals. The NRC (1971) recommended a 0.0190 per cent iodine content in salt, which is higher than their previous recommendation of 0.0076 per cent. On similar lines, the ARC (1965) recommended that dietary concentrations of iodine should be 0.8 mg/kg for pregnant and lactating animals and 0.12 mg/kg for non-pregnant and non-lactating animals when no goitrogens are fed; additional supplementation of 1.2 mg/kg is recommended with feed containing goitrogenic substances. Hemken *et al.* (1971), however, did not support the views of earlier recommendations of an additional 1.2 mg of iodine/kg to overcome goitrogenic rations.

Criteria for Iodine Deficiency Status

Iodine deficiency can be easily diagnosed when the thyroid gland is palpable or visible. Thyroid histology, iodine levels in urine and milk, serum precipitable

Table 6.9. Diagnostic data of thyroid deficiency disorders in cattle and buffaloes (ng/ml blood serum)

Species	Subjects	Concentration of thyroid hormones and PBI in serum						References
		T_4	T_3	T_3+T_4	Ratio $T_3:T_4$	%Conc. of T_3/T_4	PBI	
1. Buffaloes								
a) Neonatal calves	Male	51±5	1.0±0.1	--	64±5	--	--	Agarwal *et al.* (1989 a)
	Female	43±4	0.7±0.1	--	106±18	--	--	
b) Male calves	1-2 months	16±0.1	1.2±0.1	18±3	14±2	8±1.1	--	Agarwal *et al.* (1983 a)
	8-12 months	15±4	1.0±0.2	16±4	15±2	7±1.0	--	
	18-24 months	22±1	1.6±0.2	24±2	14±0.2	7±0.6	--	
	32-40 months	34±8	1.7±0.2	37±9	19±3	6±1.1	--	
	90-120 months	18±2	1.3±0.1	19±2	15±2	7±1.1	--	
	Overall	21±2	1.3±0.1	23±3	15±1	7±0.4	--	Dixit *et al.* (1984)
c) Bulls	10 -15 yrs. old	11±0.9	1.0±0.1	12±1	12±1	--	--	
2. Cattle								
a) Dairy cows	–	42 - 86	--	--	--	--	27-41	Kaneko (1989)
b) Bulls	Upto 48 months old	27 - 60	0.5 -1.0	55 - 70	--	--	--	Agarwal *et al.* (1983 b)

Table 6.10. Diagnostic data of thyroid deficiency disorders in sheep and goats (ng/ml blood serum)

Species	Subjects	Concentration of thyroid hormones and PBI in serum						References
		T_4	T_3	T_3+T_4	Ratio T_3:T_4	%Conc. of T_3/T_4	PBI	
3. Sheep								
a) Corriedale	Breeding season (rams)	55±2	1.1±0.03	56.41±2.2	53±2	--	--	Kaw *et al.* (1988)
	Non-breeding season (rams)	42±2	0.9±0.02	43.14±1.79	46±2	--	--	
	Breeding season (rams)	40±2	0.9±0.06	--	50±3	--	--	Agarwal *et al.* (1988)
	Non-breeding (rams)	39±2	0.6±0.06	--	66±4	--	--	
	(general)	61	--	--	--	--	38	
b) Nali	Breeding season (rams)	61±3	1.1±0.05	61.81±2.62	58±2	--	--	
	Non-breeding season (rams)	44±2	1.2±0.06	44.88±1.64	37±2	--	--	
4. Goats								
April - May	Mated	64±4	1.2±0.10	--	62±5	--	--	Agarwal *et al.* (1986 a)
	Unmated	70±2	1.0±0.10	--	90±7	--	--	
Sept. - Oct.	Mated	59±3	0.9±0.10	--	74±4	--	--	
	Unmated	69±2	1.0±0.04	--	75±3	--	--	
	(general)	53	--	--	--	--	40	Kaneko (1989)

Table 6.11. Diagnostic data of thyroid deficiency disorders in camels and pigs (ng/ml blood serum)

Species	Subjects	Concentration of thyroid hormones and PBI in serum						References
		T_4	T_3	T_3+T_4	Ratio T_3:T_4	%Conc. of T_3/T_4	PBI	
5. Camel								
a) Male	Rutting	80±4	0.4±0.0	81±4	269±45	0.5±0.1	--	Agarwal *et al.* (1986 b)
(1-4 years)	Non-rutting	109±7	1.4±0.2	110±7	79±12	1.3±0.2	--	
(4-8 years)	Rutting	78±7	0.5±0.1	79±7	202±53	0.7±0.20	--	
	Non rutting	117±11	1.3±0.1	118±11	98±8	1.1±0.1	--	
(above 8 years)	Rutting	87±4	0.2±0.1	87±4	538±244	0.2±0.10	--	
	Non-rutting	158±19	1.7±0.3	159±19	99±12	1.1±0.1	--	
b) Pregnant	Male fetus	93±3	1.1±0.0	94±3	91±4	--	--	Agarwal *et al.* (1989 b)
	Female fetus	95±3	1.1±0.3	96±2	93±3	--	--	
c) Pregnant	Upto 5 Years	92±4	1.1±0.1	94±4	91±5	--	--	Agarwal *et al.* (1989 b)
	5-10 Years	94±2	1.1±0.1	95±3	91±4	--	--	
	More than 10 years	95±3	1.1±0.1	96±3	94±4	--	--	
6. Pig								
	(general)	21.0	--	--	--	--	--	Kaneko (1989)

iodine, and protein-bound iodine are the subclinical indicators of iodine status. In the last three decades, radioimmuno assays of triiodothyronine and thyroxine have provided precise indications of iodine deficiency in various species. Doses of ^{131}I orally or i/v may be used to measure the uptake of iodine by the thyroid gland. Species-specific antibodies against the thyroid stimulating hormone (TSH) are required for better measurements of thyroid function status using R/A for TSH. Such antibodies have not been prepared for the livestock species.

Volpent and Werner (1972) reported that high levels of protein bound iodine (PBI), with a normal diet suggest adequate iodine intake. Reference values for thyroid hormones and PBI in different species of farm animals are listed in table 6.9, 6.10, 6.11; values lower than these are indicative of thyroid insufficiency. Andrewartha *et al.* (1980) reported that serum thyroxine concentration in lambs decreases with advancing age. In lactating ewes, a value of 50 n. mol/L is considered as the lower normal limit. Lambs born with congenital goitre had lower thyroxine levels than their mothers (Underwood, 1981). Rac et al (1968) reported that lambs born with congenital goitre had normal or elevated levels of thyroid hormones. It has been found that iodine deficient ewes show an increase in thyroxine concentrations after injections of iodized oil (Underwood, 1981).

Goitre should not, however, be considered as the only criterion for iodine deficiency in animals as its manifestation is dependent on species. For example, obvious enlargement of the thyroid gland is more often seen in adult sheep than in adult cattle. Clinically detectable enlargement of the thyroid gland (through inspection and palpation) is more common in goats, sheep, and young cattle than in other animals.

Work conducted on buffalo calves (Agarwal *et al.*, 1989) revealed that levels of triiodothyronine and thyroxine were high at birth, decreased considerably during the first week, and stabilized by the age of one month; peak levels occurred at puberty and maturity. Exercise might also influence the level of triiodothyronine and thyroxine in cattle bullocks where these hormones were shown to rise at the end of three hours of exercise and furthermore during eight hours of rest. This increase, however, was not statistically significant (Agarwal *et al.*, 1990).

Garg and co-workers (Garg *et al.*, 1990a,b,c) reported that in several types of reproductive disorders in livestock (i.e., anestrus, uterine torsion, difficult parturition) and in certain conditions (i.e., traumatic pericarditis, pneumonia, mastitis, fever, allergy, indigestion, tympany, impaction of rumen, intestinal obstruction, milk fever, uremia) the levels of triiodothyronine and also thyroxine were consistently lower to varying extents.

Agarwal and co-workers studied the thyroid status of camels using R/A estimations of T_3 and T_4 under varied conditions-effects of feeding a goitrogenic feed, *Leucaenao leucocephala* on growth (Rai *et al.*, 1992); following mating (Agarwal *et al.*, 1994); and after water deprivation for 20 days in winter and 10 days in summer (Deepti Khanna *et al.*, 1996).

Iodine Toxicity

In humans, it has been reported that an intake of 2000 μg of iodine/day is potentially harmful. Underwood (1977) has shown an upward trend in iodine consumption in the United States due to the cumulative intake of iodine from various sources. Iodized salt has been available in North America for many years and salt consumption is quite high.

Dairy cattle require a minimum level of 0.60 ppm of iodine in their diet and a maximum level of 50 ppm which is almost 100 times the minimum requirement (NRC, 1988). Iodine toxicity, however, may appear in dairy cattle (Hetzel *et al.*, 1987). An intake of 50 ppm of iodine in calves weighing 80 - 112 kg appears to be toxic, although some experimental animals were adversely affected at lower levels. In contrast, milking cows showed no symptoms of toxicity at iodine intakes of 50 ppm of the diet through various sources (Nai, KI), although extremely high iodine levels were present in the serum, milk, urine, and faeces (Underwood, 1977). Pigs are much more tolerant to excess iodine than cattle.

If the amount of goitrogens present in the feed is not known, then supplemental iodine should be administered to meet the minimum recommendations advocated by the NRC (1980). In this context, there is a much greater danger of iodine deficiency than iodine toxicity affecting livestock. These minimum requirements can be achieved by ensuring that the feed contains 1 per cent of iodized salt (iodized at a level of 0.01%) on concentrates and forage fed proportionately.

Conclusion

The major cause of IDD in animals, as in humans, is environmental iodine deficiency (i.e., low soil and water content reflected in iodine deficient feeds and fodder). Excessive consumption of goitrogenic substances is also a contributing factor. The effects of iodine deficiency in animals, however, are relatively less known than in human beings. Various species respond differently to iodine deficiency. It is well documented that animals living in goitre-endemic regions produce less milk, poor quality of wool and hides, and decreased reproductive performance (i.e., impaired fertility, suppressed oestrus, sterility, abortion, still births, low birth weight, and increased neonatal mortality). Regular consumption of iodine through supplementation or fortification of feeds (commercial or otherwise) has been reported to increase the productivity as well as the reproductive performance of animals.

References

NOTE: Established facts from three reviews (Miller, 1975; Underwood, 1977,1981) have been used as baseline data. Most of the original sources of these facts are not mentioned in the bibliography. Only additional and more recent publications have been listed.

Agarwal, S.P., A.K., Agarwal, N., Singh, P.K., Dwarkanath. 1983a. 'Serum testosterone and thyroid hormone levels in male buffalo-calves of different ages'. *Indian J. Anim. Sci.* 53: 609-11.

Agarwal, V.K., S.P., Agarwal, N., Singh, P.K., Dwarkanath. 1983b. 'Levels of serum thyroid hormones in relation to age and sexual development of cross-bred bulls'. *Indian J. Anim. Sci.* 1983b. 53: 1063-65.

Agarwal, S.P., V.K., Agarwal, A.S., Kanaujia, P.K., Dwarkanath. 1986a. 'Circulating levels of thyroid hormones in mated and unmated goats during estrous cycle'. *J. Nuclear Argric. Biol.* 15: 86-9.

Agarwal, S.P., N.D., Khanna, V.K., Agarwal, P.K., Dwarkanath. 1986b. 'Thyroidal status of male camel during breeding and non-breeding seasons'. *Indian J. Anim. Sci.* 56: 1036-38.

Agarwal, V.K., S.L., Garg, B., Vasudevan, S.P., Agarwal. 1988. 'Seasonal variations in peripheral testosterone, cortisal and thyroid hormones in Corriedale rams under tropical climate'. *J. Nuclear Agric. Biol.* 17: 257-59.

Agarwal, S.P., M.S., Saini, B.P., Sengupta, V.K., Agarwal. 1989a. 'Thyroidal status of neonatal and young buffalo calves'. *J. Nuclear Agric. Biol.* 18: 227-32.

Agarwal, S.P., N.D., Khanna, V.K., Agarwal, P.K., Dwarkanath. 1989b. 'Circulating concentrations of thyroid hormones in pregnant camels (Camelus dromedarius)'. *Theriogenology*. 31: 1239-47.

Agarwal, S.P. et al., 1994. Thyroid Status of female Camels following mating. *Indian Vet.J. 72, 591.*

Andrewartha, K.A. et al. 1980. *Aust. Vet. J.* 56: 18.

ARC. 1965. *The Nutrient Requirement of Farm Livestock.* No.2. Agriculture Research Council London: U.K.

Barry, T.N., S.J., Duncan, W.A., Sadler, K.R., Tillar, A.D., Sheppard. 1983. 'Iodine metabolism and thyroid hormone relationships in growing sheep fed on kale (Brassica oleracea) and ryegrass (Lolicon perenne) - Clover (Trifolium repens) fresh forage diets'. *British Journal of Nutrition.* 49: 241-53.

Bedi, S.P.S. IANWC for Asia and Pacific, Bangalore 191 Compendium II (Abstract). p. 76.

Berg, J.N., D., Padgitt, B., Mc Carthy. 1988. 'Iodine concentrations in milk of dairy cattle fed various amounts of iodine as ethlenediamine dihydroiodid'. *Journal of Dairy Science.* 71: 3283-91.

Bodal, J. 1976. 'The effect of phosphorus and trace element deficiency on reproduction and newborn calves'. *Magyar Allatorvosok Lapja.* 31: 589-93.

Budmaska, V. Yu, N.I., Terpelyuk, A.A., Demidyuk. 1983. 'Milk yield of cow given trace element supplements'. *Zhivotnovodstvo*. 3: 54-5.

Cole, V.V. 1935. *J. Nutr.* 10: 493.

Conrad, I.M. III., R.W., Hemken.1978. ' Milk iodine as influenced by an iodophor teat dip'. *J. Dairy Sci.* 61: 776.

Cromwell, G.L., D.T.H., Sihombing, V.W., Hays. 1975. 'Effects of iodine level on performance and thyroid traits of growing pigs'. *Journal of Animal Science*. 41: 813-818.

Curtis, G.M., I.D., Puppel, V.V., Cole, N.L., Matthews. 1937. *J. Lab. Clin. Med.* 22: 1014. In: Underwood, E.J. *Trace Elements in Human and Animal Nutrition*. 4th ed. Academic Press: New York. 1977.

Deepti Khanna, S.P., Agarwal, M.L., Gupta, A.K., Rai and N.D., Khanna. 1996. 'Effect of water deprivation during summer and winter on thyroid hormones concentration in the Indian camel'. *Indian J. Anim. Sci. 66, 253-55,*

Devilat, J., A., Skoknic. 1971. 'Feeding high levels of rapeseed meal to pregnant gilts'. *Canadian Journal of Animal Scinece*. 51: 715-19.

Dixit, N.K., S.P., Agarwal, V.K., Agarwal,. 1984. 'Seasonal variations in serum levels of thyroid hormones and their relation with seminal quality and libido in buffalo bulls'. Theriogenology. 22: 497-507.

Dunsmore, D.G. 1976. 'Iodophors and iodine in dairy products. The iodine contents of Australian dairy products'. *Australian J. Dairy Technol.* 31: 125.

Dunsmore, D.G., C., Nuzum, B., Dettmen. 1977. 'Iodophors and iodine in dairy products: 3 teat dipping'. *Australian J. Dairy Technol.* 32: 45.

Elmer, A.W. *Iodine Metabolism and Thyroid Function*. Oxford University Press: London and New York. 1938. Falconer, I.R. 1987. Comp. Pathol. Bull. 19: 2.

Fisher, K.D., C.J., Carr. *Iodine in Foods: Chemical Methodology and Sources of Iodine in the Human Diet*. Life Sciences Research Office. FASEB: Bethesda. 1974. MD (FDA 71-294).

Garg, S.L., N.S., Bugalia, Suresh Chander, A.R., Rao. 1989a. 'Plasma tri-iodothyronine (T_3) profile in anoestrus and cycling cows'. Indian Soc. Reproduction at Hisar.

Garg, S.L., S., Chander, A.R., Rao. 1989b. 'Peripheral levels of thyroid hormones in periparturient buffaloes'. Soc. Anim. Physiologists India. Matpatty, India.

Garg, S.L., N.S., Bugalia, S., Chander, A.R., Rao. 1990a. 'Thyroid hormones in buffaloes in anoestrus treated with Norgestomet'. Congress of Asian and Oceanian Physiological Societies. New Delhi.

Garg, S.L., N.S., Bugalia, S., Chander, A.R., Rao. 1990b. 'Plasma tri-iodothyronine and tetra-iodothyronine (T_4) levels in buffaloes in anoestrus'. Soc. Anim. Physiologists India. Khanapara, India.

Garg, S.L., A.R., Rao. 1990c. 'Thyroid hormones in crossbred calves suffering from foot and mouth disease'. Indian Soc. Vet. Med. Hyderabad, India.

Gogawale, M.N. and M. Singh (1984). Mineral status of animals as influeneed by geneties and geography. *In Proceedings of the Symposium on recent advances in mineral nutrition,* 'Haryana Agricultural University', Hisar.

Golubev, M.I., S.P., Sedov. 1989. *Effect of Iodine on Liveweight and Milk Production in Cattle.* Gor'kii, USSR. pp. 16-19.

Gridin, V.F. 1985. 'Influence of iodine and cobalt on digestion in the rumen of lactating cows and their productivity'. Biokhimii i Pitaniya Sel'skokhozyaistvennykh Zhivotnykh. 30: 74-8.

Groppel, B., M., Anke. 1983a. 'Diagnosis of iodine level'. Spurenelement - Symposium. 4: 156-63.

Groppel, B., R., Korber, M., Anke, A., Henning. 1983b. 'Iodine deficiency in goats, sheep and cattle'. Spurenelement Symposium. 4: 164-70.

Groppel, B., M., Anke, B., Kohler, E., Scholz. 1989. 'Iodine deficiency in ruminants. 1. The iodine content of feedstuffs, plants and drinking water'. Archives of Animal Nutrition. 39: 211-20.

Healy, W.B., G., Croucheley, R.L., Gillett, P.C., Rankin, H., Watis. 1972. 'Ingested soil and iodine deficiency in lambs'. *New Zealand Journal of Agricultural Research.* 15: 778-82.

Heinrich, H., E.E., Gabb, D.H., Whang. 1964. Atomkernenergie. 9: 279.

Heinrich, H., G., Wenk, G., Bugdol. 1989a. 'Effect of iodising mineral mix for cattle, sheep and pigs on iodine supply in man'. In: *6th International Trace Element Symposium*. 1989. 3: 756-62.

Heinrich, H., G., Wenk. 1989b. 'Iodine content of feeds and water used for cattle, sheep and pigs in the Erfurt area'. In: *6th International Trace Element Symposium*. 3: 749-55.

Hemken, R.W. 1970. *J. Dairy Sci*. 53: 1138-43.

Hemken, R.W., J.H., Vandersall, B.A., Sass, J.W., Hibbs. 1971. 'Goitrogenic effects of a corn silage-soybean meal supplemented ration'. *J. Dairy Sci*. 54:85.

Hemken, R.W., J.H., Vandersall, M.A., Oskarsson, L.R., Fryman. 1972. 'Iodine intake related to milk iodine and performance of dairy cattle'. *Journal of Dairy Science*. 55: 931-34.

Hetzel, B.S., M.T., Mano. 1989a. 'A review of experimental studies of iodine deficiency during fetal development'. *Journal of Nutrition*. 119: 145-51.

Hetzel, B.S., M.T., Mano, D.A., Roland. 1989b. 'Iodine deficiency in young male cattle'. Veterinary (Mascow, USSR). No. 15: 55-6.

Iwarsson, K., G., Bengtsson, L., Ekman. 1973. ' Iodine content in colostrum and milk of cows and sows'. Acta Veterinaria Scandinavica. 14: 254-62.

Iwarrson, K., L., Ekman. 1974. ' Iodophor teat dipping and the iodine concentration of milk'. *Nord. Vet. Med*. 26:31.

Joerin, M.M., A., Bowering. 1972. 'Total iodine content of cow's milk'. New Zealand *J. Dairy Sci. Technol*. 7: 155. Kaneko, J.J. 'Thyroid function'. In: Felig, P., J.D., Baxter, A.E., Broadus, L.A., Frohman.(eds.). *Endocrinology and Metabolism*. McGraw-Hill: New York. pp. 630-48.

Karmarkar, M.G., M.G., Deo, N., Kochupillai, V., Ramalingaswami. 1974. 'Pathophysiology and Himalayan endemic goiter'. *The American Journal of Clinical Nutrition*. 27: 96-103.

Kaw, V.C., S.P., Agarwal, A.R., Rao. 1988. 'Thyroid hormone levels in sheep'. *Indian Soc. for Nuclear Techniques in Agric. and Biology*. New Delhi.

King, C.F. 1976. *Aust. J. Exp. Ag. an. Hios.* 16: 651.

Knights, G.I., P.K., O'Rourke, P.S., Hopkins. 1979. 'Effects of iodine supplementation of pregnant and lactating sows on the growth and maturation of their off springs'. *Australian Journal of Agricultural and Animal Husbandry.* 19: 19-22.

Kobeisy, M.A., E.M., Ibrahim, M.M., Shetaewi. 1992a. 'Composition and some technological properties of water buffalo milk as affected by dietary iodine supplementation'. *Assiut Veterinary Medical Journal.* 27: 111-22.

Kobeisy, M.A., M.M., Shetaewi. 1992b. 'Effect of supplemental dietary iodine on thyroid hormone, serum constituents and performance of lactating water buffaloes'. *Assiut Veterinary Medical Journal.* 28: 129-40.

Koiranen, L., R., Stabel-Taucher. 1976. 'Iodine content of Finnish milk'. *Suomen Elainlaakarilehti.* 82: 545-49.

Korber, R., H., Gurtler, G., Pethes, G., Furcht, R., Wenzel, P., Rudas, W., Spielke. 1983. 'Thyroid function in iodine deficient piglets, and oral iodine therapy for pregnant and lactating sows'. Monatshefte fur Veterinarmedizin. 38: 694-99.

Korber, R., N., Rossow, J., Otta. 1985. 'Iodine deficiency syndrome in cattle, sheep and swine'. Monatshefte fiir Veterinarmedizin 1985. 40: 220-24.

Koval'kova, R., M., Selikova, N., Metelina. 1984. 'Effect of potassium iodide and sodium sulphate on productivity of cows'. Molochnoe i Myansoe Skotovodstvo. 4:25.

Koval'skil, V.V., A.I., Ladan, I.F., Gribovskaya, R.I., Blokhina. 1972. 'The amur biogeochemical region of the iodine deficiency'. Vestnik Sel'skokhozyaistvennoi. 10: 66-70.

Koval'skil, V.V., I.F., Mavrin. 1974. ' Increasing productivity of cattle given potassium iodide supplement'. Vestnik Sel'skokhozy aistvennol Nauki. 10: 65-70.

Levine, H., R.E., Remington, H.V., Kolnitz. 1933. 'Studies on the relation of diet to goitre 2. The iodine requirement of the rat'. *J. Nutr.* 6: 347-54.

Mason, R.W. 1976. 'Milk iodine content as an estimate of the dietary iodine status of sheep'. *British Veterinary Journal.* 132: 374-79.

Mavrin, I.F., V.V., Koval'skill, A.I., Ladan. 1976. 'Effectiveness of iodine supplements in animals husbandry of the Zeya-Burein plain'. Sel'skokhozyaistvennykh Nauk. 2: 32-4.

McGrath, D., G.A., Fleming. 1988. 'Iodine levels in Irish soil and grasses'. *Irish Journal of Agricultural Research.* 27: 75-81.

Miller, W. J. *Dairy Cattle Feeding and Nutrition.* Academic Press: New York. 1975.

Myberg, J.A. 1970. Acta Vet. Scand. Suppl. 31:120.

Newton, G.L., A.J., Clawson. 1974. *J. Anim. Sci.* 39:879.

NRC. 1971. *Nutrient Requirement of Dairy Animals.* 4th ed. 'National Academy of Sciences: Washington', D.C.

NRC. 1978. *Nutrient Requirement of Dairy Animals.* 5th ed. 'National Academy of Sciences: Washington', D.C.

NRC. 1980. *Nutrient Requirement of Dairy Animals.* 'National Academy of Sciences: Washington', D.C.

NRC. 1988. *Nutrient Requirement of Dairy Animals.* 6th ed. (revised). 'National Academy of Sciences: Washington', D.C.

Ogoleva, V.P., N.K., Besserezhnova, A., Lukshin, G.T., Kovaleva. 1987. 'Iodine in animal production in lower Volga'. Khimiya V Sel'skom Khozyaistve. 25: 30-3.

Paxman, P.J., R.J., Hill. 1974. *Sci. Food Agric.* 25: 323.

Pyska, H. 1974. 'Iodine and thiocyanate in the milk of cows at different stages of lactation'. Roczniki Naukowe Zootechniki.1: 73-9.

Rac, R., G.N., Hill, R.W., Pain, C.J., Mulhearn. 1968. *Res. Vet. Sci.* 9: 209.

Rai, A.K. et al. *1992.* 'Effect of feeding Leucaena leucocephala with Phaseolus aconitofolius on growth and thyroid status of camel calves'. *Indian J. Anim. Sci. 62, 297.*

Rudent, C.P., J., Oliver. 1976. *Rhod, J. Agric. Res.* 14:67.

Ryot, K.D., B.K., Sharma, C.D., Panwar. 1990. 'Effect of iodine therapy in anoestrous bovines'. *Indian Journal of Animal Production.* 11: 144-45.

Salakhutdinov,K.G. 1985. *Iodine Deficiency in Farm Animals.* Kazanskii Veterinarnyt Institut: Kazan, USSR. pp. 96-99.

S'edina, G.V. 1987. 'Effect of iodine nutrition of cows on their milk yield and biochemical values of blood'. Byulletin Vsesoyuznogo Nauchno-issledovatel's Kogo Insatituta Razvedeniyai Genetiki Sel'skokhozyaistvennykh Zhivotnykh. 100: 17-9.

Scoffer, L. 1933. Biochem. Z. 259: 11. In: Underwood, E.J. *Trace Elements in Human and Animal Nutrition.* 4th ed. Academic Press: New York. 1977.

Scott, M.L., D.A., Van Tien Loven, E.R., Holm, R.E.J., Reynolds. 1960. Nutr. 70:282.

Shuman, A.C., D.P.J., Townsend. 1963. Anim. Sci. 22:72. In: Understood, E.J. *Trace Elements in Human and Animal Nutrition.* 4th ed. Academic Press: New York. 1977.

Svezhentosov, A.I., S.I., Toma, E.P., Kaptenko. 1975. ' Contents of iodine in feeds and plants in the Moldavian SSR'. Khimiya V Sel'skom Khozyaistve. 13: 919-921.

Talbot, J.M., K.D., Fisher, C.J., Carr. *A Review of the Significance of Untoward Reactions of Iodine in Foods.* 'Life Sciences Research Office'. FASEB. Bethesda. 1974. MD (FDA 71-294).

Talbot, J.M., K.D., Fisher, C.J., Carr. *A Review of the Effects of Dietary Iodine on Certain Thyroid Disorders.* 'Life Sciences Research Office'. FASEB. Bethesda. 1976. MD (FDA 71-294).

Timoshenko, N. 1975. ' Generative function of the ovaries in sows given supplements of potassium iodide and kormogrizn'. Svinovodstvo (Moscow, USSR). 8:38.

Underwood, E.J.' The Mineral Nutrition of Livestock'. 1st ed. CAB: England. 1981.

Unsi-Ranva, E., E., Tanhuanpac, A., Unsi-Ranva. 1974. 'Transfer of iodine from teat dip into milk and udder tissues'. Supm. Elain 79:729.

Volpert, E.M., S.C., Werner. 1972. 'Serum tri-iodothyronine concentration in the iodine deficient rats'. *American Journal of Anatomy*. 135: 187-90.

Wayne, E.J., D.A., Koutras, W.D., Alexander. *Clinical Aspects of Iodine Metabolism*. Blackwell: Oxford. 1964.

Wilghs, H.S. Jr., E.X., Gassner, A.R., Patton, G.S., Harshfield. 1953. 'The iodine requirement of chickens'. Technical Bulletin. Colorado Agricultural Experiment Station. No. 49.

Wood, F.O. In: *Iodine Nutrition in the United States*. Natl. Acad. Sci. Natl. Res. Counc., Washington, D.C. 1970. p.30.

Zamarin, L.G., A.V., Minullin, R.N., Akhmadeev. 1990. 'Influence of iodine and cobalt on carbohydrate and fat metabolism in cows'. Veterinariya (Hoskva). 2: 55-8.

7

Effects of Iodine Deficiency on Livestock

S.P. Pachauri

Abstract

Iodine is an essential element in maintaining the general homeostasis in the animal body through thyroid hormones and its deficiency can affect many physiological systems. A naturally occurring deficiency of iodine in the soil, water, and feed of livestock is the primary cause of low iodine intake. Goitrogenic substances found in certain feed crop species which inhibit iodine uptake in the thyroid gland are secondary factors in precipitating iodine deficiency disorders in livestock. The specific effects on iodine deficiency in cattle, sheep and goats, pigs, horses, and poultry are described and the prophylactic administration of iodine in salt is briefly discussed.

Introduction

The cardinal sign of iodine deficiency in livestock is goitre, which is the final expression of a number of pathological disorders associated with the shortage of iodine in the animal body. The mature animal body is estimated to contain less than 0.00004 per cent of iodine, but if this minute amount is not maintained through the diet, disaster results. Iodine is an essential element, the concentration of which must be maintained within quite narrow limits or normal ranges if the growth, health, and productivity of the animal are to remain unimpaired. Continued ingestion of diets deficient in iodine, induces changes in the concentration of this mineral in the body tissues and fluids resulting in biochemical lesions and adverse physiological functions. About 80 per cent of the body iodine is stored in the thyroid gland in the form of iodothyroglobulins, thyroxine, triiodothyronine, and trace quantities of other iodinated amino acid derivatives. Iodine apparently exerts its total physiologic role as a component of the thyroid hormones, which control cellular energy exchange and basal metabolic rate, as well as tissue growth, reproduction, lactation, and development.

Simple iodine deficiency results in an enlargement of the thyroid gland (goitre) with a reduction in thyroxine secretion. Hyperplastic changes in the thyroid begin as a result of the failure of thyroid tissue to supply enough secretion, due to either a reduced supply of iodine for the manufacture of thyroxine or due to an increased demand of the hormone by the body.

Factors Causing Iodine Deficiency

Iodine deficiency is primarily due to deficient iodine intake as a consequence of low levels of available iodine in the soil. A high intake of calcium, gross bacterial pollution of food stuffs or drinking water, and the presence of goitrogenic substances, which interfere with the utilization of iodine by the thyroid, are secondary factors.

Primary Factor: Deficient Iodine Intake

The recommended dietary intake of iodine is 0.1- 0.3 mg/kg dry matter of food (DM). The iodine requirements of livestock are sometimes not met in the natural diet and iodine deficiency occurs. The demand for thyroxine varies with the activity of the body functions that it controls and thus, given a constant supply of iodine in the diet, simple goitre is most likely to develop during periods of greatly increased need. In humans, these critical periods occur during pregnancy and puberty. In farm animals, goitre usually shows itself in the young at birth as a result of iodine deficiency in the rations of the mother during gestation. The affected young are born weak or dead. On a deficient diet, the pregnant animal is not able to supply the fetus with enough iodine. The danger increases in the case of multiple births.

Iodine deficiency occurs in areas in which soil (and hence plant and water) iodine content is low (Pachauri, 1981). Interest in soil analysis as a means of assessing the availability of iodine has been minimal. Values for total iodine have been used as an index of the iodine status of food, plants, and water in an effort to delineate goitrous areas (Raina, 1981). This can only be useful if viewed against the effects of various iodine enriching agencies (i.e., atmospheric precipitation, vegetation, clay content of soils, and acidity) and depleting agencies (i.e., leaching, alkalinity, and cropping).

Soil levels of iodine were studied (Pachauri, 1981) and Raina and Pachauri (1984) suggested a correlation between goitre and soil levels of iodine (table 7.1). These observations led to the conclusion that iodine content of Terai soil should be much higher than the levels recorded because at higher pH values, iodine is presumably displaced from clay surfaces by hydroxyl ions; and with increased acidity of clay suspensions, there is a progressive increase in iodine absorption, as reported by Hamid and Warkentin (1967).

Secondary factors: Goitrogenic Substances

Although a lack of iodine is the primary cause of simple goitre, it is recognized that other factors may contribute, notably the high calcium content of soil and water in many goitrous regions. In addition, there are some plants, such as cabbage, peanuts, soybeans, yellow turnips, members of the brassica family, and cruciferous crops that generally may promote goitre development as they contain substances (goitrogens) which inhibit thyroxine production. The active component of these plants/foods is glucoside. A continued intake of a low level of cyanogenic glycosides (e.g., white clover) is commonly associated with a high incidence of goitrous offspring. Linamarin, a glucoside in linseed meal, is thought to be the agent producing goitre in new born lambs from ewes fed with the meal during pregnancy. A continued intake of the grass *Cynoden aethiopious* with low iodine and high cyanogenic glucoside content has also been found as a cause of goitre in lambs (Rubert and Oliver, 1976). Rape seed and rape seed meal are also goitrogenic. Korber *et al* .(1985) reported a high prevalence of iodine deficiency in cattle, sheep, and swine herds fed diets containing nitrates, thiocyanates or glucosinolates.

A high dietary intake of calcium also decreases intestinal absorption and in some areas, heavy applications of lime to pastures are followed by the development of goitre in lambs. This factor may also be important in areas where drinking water is heavily mineralized.

Table 7.1. Iodine content of soil collected from the Terai region in Uttar Pradesh, India

Place	Texture	pH	Iodine (ppm)
Bhimtal	Loam	6.4	7.41
Bhowali	Loam	6.9	6.41
Kathgodam	Loam	7.1	6.23
Haldwani	Loam	6.8	7.24
Rudrapur	Loam	7.9	8.38
Pantnagar	Loam	8.1	5.65
Pantnagar	Silt loam	6.5	5.90
Kichha	Silt loam	7.5	11.49
Bilaspur	Silt loam	7.7	8.60
Sultanpurpatti	Silt loam	7.5	7.98
Kelaphera	Silt loam	7.5	5.90
Lalkuan	Sandy loam	6.8	10.93
Pantnagar	Sandy loam	7.8	5.51
Nainital	Sandy loam	5.5	6.66
Kashipur	Clay loam	7.0	6.87
Gadarpur	Clay loam	7.8	6.53
Pantnagar	Clay loam	7.7	6.06

Under experimental conditions, groundnuts are goitrogenic for rats, the goitrogenic substance being a glycoside, arachidoside. Soyabean by-products are also considered to be goitrogenic. Gross bacterial contamination of drinking water by sewage is a cause of goitre in humans in countries where hygiene is poor (Blood *et al.* 1989). There is a record of a severe outbreak of goitre in calves from cattle running on pasture heavily dressed with crude sewage. Feeding sewage sludge is also linked to the occurrence of goitre.

Effects of Iodine Deficiency

Iodine deficiency results in a decreased production of thyroxine and stimulation of the secretion of thyrotropin hormone by the pituitary gland. This commonly results in hyperplasia of thyroid tissue and a considerable enlargement of the gland. Most cases of goitre of the newborn are of this type. The primary deficiency of thyroxine is responsible for the severe weakness and hair abnormality of the affected animals. Although the defect is described as hairlessness, it is more properly described as a hypoplasia of the hair, with many very slender hairs present and a concurrent absence and diminution in size of hair follicles (Itacura *et al.* 1979). A hyperplastic goitre is highly vascular, the gland can be felt to pulsate with an arterial pulse, and a loud murmur may be audible over the gland. Colloid goitre is less common in animals and probably evolues after primary hyperplasia.

Other factors, particularly the ingestion of low levels of cyanide, probably exert their effects by

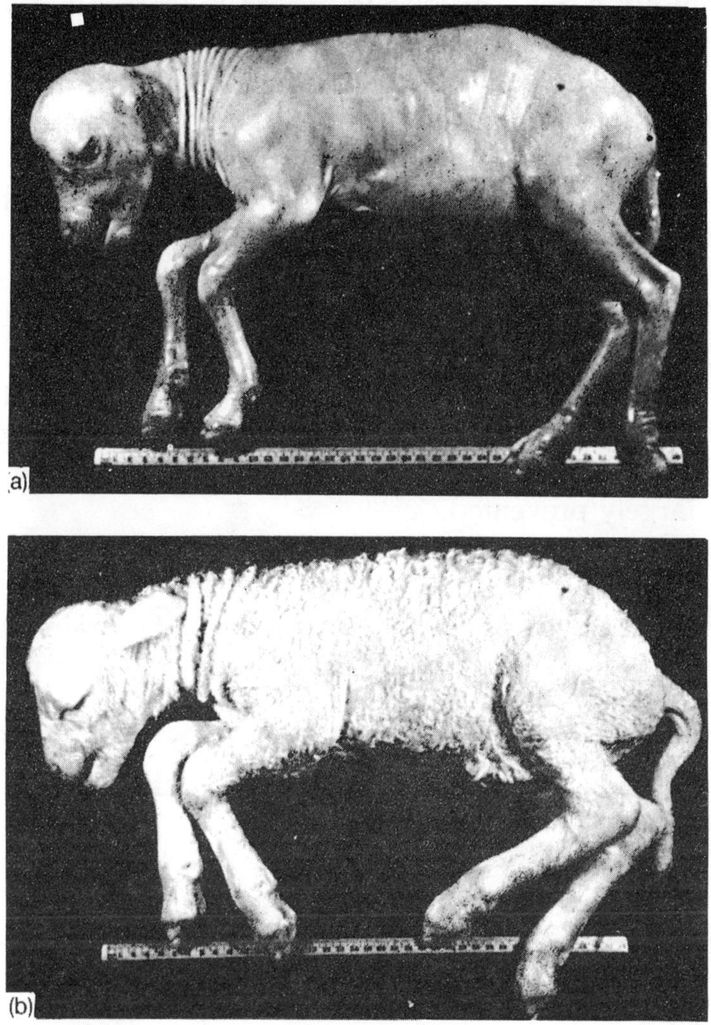

Fig. 7.1. Effect of severe iodine deficiency during pregnancy on the development of lamb. A 140-day-lamb foetus (a) was subjected to severe iodine deficiency by feeding the mother an iodine deficient diet for 6 months prior to and during pregnancy (full term 150 days), compared to a control lamb (b) of the same age fed the same diet but with the addition of an iodine supplement. The iodine deficient lamb showed an absence of a woolly coat, dislocation of the leg joints, with a smaller brain. The figure illustrates that iodine is essential for animal development and is important to the animal industry (From Hetzel 1989 with permission).

inhibiting the metabolic activity of the thyroid epithelium and restricting the uptake of iodine. Thiocyanates and sulfocyanates are formed during the detoxification of cyanide in the liver and these substances have a pronounced depressing effect on iodine uptake by the thyroid. Some pasture and fodder plants, including white clover, rape, and kale are known to have a moderate content of cyanogenic glucosides. These goitrogenic substances may appear in the milk and provide a toxic hazard to both animals and man. The inherited form in cattle appears to be due to the increased activity of an enzyme which de-iodinates iodotyrosines so rapidly that the formation of thyroxine is largely prevented.

Iodine is an essential element for fetal brain development in sheep. A severe iodine deficiency in pregnant ewes causes a reduction in fetal brain and body weight from 70 days of gestation to parturition (Blood *et al.* 1989).

Manifestations of Iodine Deficiency in Animals

The manifestation of iodine deficiency in cattle, sheep, goats, pigs, horses, and poultry are described below.

Cattle

Clinical manifestations of iodine deficiency in cattle are a gross enlargement of the thyroid gland and weakness in newborn calves. Morbidity is high, especially shortly after birth. If calves are not assisted to suck for a few days they may die. In some instances, the thyroid gland is sufficiently large to cause obstruction to respiration. Partial alopecia is a rare accompaniment. Loss of libido in bulls, a failure to express estrus in the cow, and a high incidence of aborted, still born or weak calves have been suggested as manifestations of hypothyroidism in cattle.

Sheep and goats

Adult sheep in iodine deficient areas may show a high prevalence of thyroid enlargement but are clinically normal in other respects. Newborn lambs manifest weakness, extensive alopecia, and palpable, if not visible, enlargement of thyroid glands. Prolonged gestation is reported in ewes.

Goats present a similar clinical picture, except that all abnormalities are more severe than in lambs. Thyroid glands of goats/kids may be graded as palpable and plum size (+), easily palpable and lemon size (++), and duck egg size, hanging and visible from a distance (+++). Occurrence of goitre varied between 0.5 per cent and 5.5 per cent at Laxmipur goat farm in Deoria, Uttar Pradesh (Rajkumar, 1970) and 0.16 - 5.66 per cent in the Terai region of Uttar Pradesh (Raina and Pachauri, 1984). The prevalence in kids is higher than in adults and the degree of enlargement is more in kids than in their dams (Pachauri, 1981). Animals which survive the initial danger period after birth may recover except for partial persistence of the goitre. The glands may pulsate with the normal arterial pulse and may extend down a greater part of the neck and cause some local edema.

Auscultation and palpation of the jugular furrow may reveal the presence of a murmur and thrill (the 'thyroid thrill') due to the increased arterial blood supply of the glands.

Pigs

In pigs, the major findings are the birth of hairless, still born, or weak piglets often with myxedema of the skin of the neck. Hairlessness is most marked on the limbs. Most affected piglets die within a few hours of birth. Thyroid enlargement may be present but it is never sufficiently large enough to cause visible swelling in the live pig. Survivors are lethargic, do not grow well, have a waddling gait, and have leg weaknesses due to weakness of ligaments and joints.

Horses

Foals show a normal hair coat and little thyroid enlargement but are very weak at birth. In most cases, they are unable to stand without support and many are too weak to suck. Excessive flexion of the lower forelegs and extension of lower parts of the hindlegs has also been observed in affected foals. Defective ossification has also been reported, the manifestation being the collapse of the central and the third tarsal bones leading to lameness and deformity of the hock. Enlargement of the thyroid also commonly occurs in adult horses in affected areas with thoroughbreds and light horses being more susceptible than draft animals.

Poultry

Iodine deficiency in poultry results in an enlarged thyroid and in some cases lower body weight in growing chicks. Wilgus *et al.* (1941) observed congenital goitre in baby chicks hatched from hens receiving 0.025 ppm iodine in the ration. Rogler *et al.* (1959) observed mortality late in incubation. Hatching time was delayed. Embryo size was reduced and yolk-sac resorption was retarded.

Salt and Iodine

Iodine can be provided as a fertilizer, in salt, or in a mineral mixture. Since salt requirements must also be met, it is practical to provide iodine in common salt. Requirement of sodium is 53 to 64 mg/kg of body weight, which would be provided by 0.3 per cent of salt in the dry matter of the ration. The supplemented salt required for a lactating cow producing 16 kg of milk daily is estimated to be about 30 g per day, in addition to that used in feeds. In the case of pigs and chickens, the requirement of salt is 0.25 to 0.5 per cent in the dry ration. It is commonly recommended that lambs and calves have free access to salt, however, the practice of allowing salted *libitum* to farm stock undoubtedly results in an intake in excess of requirements.

The recommended dietary intake of iodine for cattle is 0.8 - 1.0 mg/kg dry matter of feed for lactating and pregnant cows and 0.1 to 0.3 mg/kg dry matter of feed for empty cows and calves. The iodine requirement of pregnant cows is approximately 0.44 mg/100 kg of body weight daily. Growing swine require somewhat less than this. Adult swine require 0.20 mg iodine per kg of diet.

For chicks, the requirement has been set to 0.35 mg/kg of feed. These levels for chicks are higher than those found adequate for goitre prevention in other species. One reason for this high recommendation is to provide insurance against the goitrogenic effect of soybean meal which is commonly used as a feed ingredient.

It is a common recommendation that cattle, sheep, and horses receive 1 per cent of stabilized iodised salt, containing 0.0076 per cent iodine in the grain portion of their rations and that swine be fed this salt as 0.5 per cent of the grain ration.

Use of 0.25 per cent iodized salt in chicken and turkey rations prevents development of iodine deficiency. This would supply 0.175 ppm in addition to that contained already in the diet. Christensen (1988) reported that dietary supplements of iodine increased the permeability of egg shells and the hatchability of turkey eggs.

Conclusion

The effects of environmental iodine deficiency, and to some degree goitrogens, on various livestock species pose a serious threat to the farmer as animal health and productivity can be seriously affected. Iodized salt should be administered on a prophylactic basis to those animals in affected regions in measured amounts on a consistent basis so as to prevent iodine deficiency disorders in livestock.

References

Blood, D.C. and O.M., Radostits. 1989. *Veterinary Medicine : A text book of the diseases of cattle, sheep, pigs, goats and horses*. 7th ed. ELBS Bailliere Tindall: U.K. pp. 1176.

Christensen, V.L. and J.F., Orf. 1988. 'Effect of dietary iodine on the permeability and hatchability of large white turkey eggs'. *Poultry Science*. 67: (Suppl).

Hamid, A., B.P., Warkentin. 1967. 'Retention of ^{131}I used as a tracer in water movement studies'. *Soil Scientist*. 104: 279-82.

Itakaru, C., D., Nakano and M., Goto. 1979. 'Histopathologic features of the skin of hairless newborn pigs with goitre'. *Amer. J. Vet. Res*. 40: 279-82.

Korber, R., N., Rossow and J., Otta. 1985. 'Iodine deficiency syndrome in cattle, sheep and swine'. Monatshefte fur Veterinar Medizin. 40: 220-24.

Pachauri, S.P., 1981. 'Clinical studies on endemic goitre in goats in Tarai'. Status report, Department of Medicine, College of Veterinary Sciences, G.B. Pant University of Agriculture and Technology, Pantnagar, Dist. Nainital.

Raina, A.K. 1981. 'Clinical studies on endemic goitre in goats in Tarai'. M.V. Sc. Thesis. Department of Medicine, College of Veterinary Sciences, G.B. Pant University of Agriculture and Technology, Pantnagar, Distt. Nainital.

Raina, A.K. and S.P., Pachauri. 1984. 'Studies on prevalence of goitre in goats in Tarai'. *Indian Vet. J*. 61: 684-88.

Rajkumar, S.S., 1970. 'Incidence of goitre in goats'. *Indian Vet. J*. 47: 185-7.

Rogler J.C., H.E., Parker, F.N., Andrews, C.W., Carrick. 1959. 'The effects of an iodine deficiency on embryo development and hatchability'. Poultry Science. 38: 398-405.

Rudart, C.P., J., Oliver. 1976. 'Effect of thiocyanate on the occurrence of goitre in new born lambs'. *Rhod. J. Agric. Res*. 14: 67-72.

Wilgus, H.S. Jr., Harshfield., G.S., Patton, A.R., Ferris, L.P., Gassner. 1941. 'The iodine requirements of growing chickens'. Poultry Science. 20: 477.

8

Iodine Deficiency and Animal Production

James A. Pittman, (Jr.) and Constance S. Pittman

The essential role of iodine in the nutrition of animals has attracted less attention among physicians and the public at large than has iodine deficiency in human beings, because the latter affects us all directly and dramatically —or has the danger of doing so, while iodine deficiency in animals affects human beings less directly and appears to us important more from an economic perspective. Deficiency of iodine or other micronutrients can make the animals ill and lead to important economic losses.

For veterinary doctors and farmers, as for physicians attending human patients, the problems are threefold: Dx, Px, and Rx. That is, diagnosis, prognosis, and therapy. As in human medicine, the term 'diagnosis' implies far more than just giving a condition a name. It comes from two Greek roots, 'dia,' which means 'through' or 'thorough,' and 'gnosis,' or 'knowledge.' Thus, to make a diagnosis implies obtaining a thorough and complete understanding of the entire disease process, from etiology, through pathogenesis, through mechanisms of production of symptoms and signs, through the clinical manifestations themselves, to knowledge of the likely future course of the disorder and possible remedies for it.

In considering diseases of animals, particularly nutritional disorders, we lose the subjective symptoms, which are unavailable to help with making a diagnosis. We are left with only the clinical manifestations of disorders which we can ourselves observe – that is, the signs of the disease. However, veterinarians recommend a clinical work-up quite similar to that of a physician considering human ailments. That is, 'The clinician who is confronted with a problem where trace minerals are suspected should always obtain a detailed herd history, perform a complete physical examination on affected animals, evaluate the total diet, including pastures, and harvest tissues needed to evaluate the specific mineral concentrations in the body' (Pugh, 1995).

As with human diseases, the differential diagnosis is important; that is, differentiating the presumed diagnosis from other diagnostic possibilities and ruling out the latter, leaving the observer with the final diagnosis. This is important because animals suffering from iodine deficiency may also suffer from deficiency of other essential nutrients, and these must not be overlooked. This is considered in many textbooks on veterinary medicine. However, our chief concern here is with iodine.

The main manifestations of iodine deficiency are tabulated in table 8.1, which includes effects in many species.

There is a very wide variety of animals used by human beings for a wide variety of purposes. The list shown in table 1 was derived from those published for common farm animals around the world, but it probably applies to all animals. There is no reason to think that this list of clinical manifestations of iodine deficiency would be different in different parts of the world, through

Table 8.1. IDD manifestations in animals

Reproduction ('Reproductive failure')	
	Decreased reproductive frequency (mating)
	Decreased male libido
	Decreased fertility
	Decreased egg production
	Decreased hatchability
	Goitre and death of embryo
	Increased fetal resorption
	Increased abortion
	Prolonged gestation
	Increased stillbirths
	Retention of fetal membranes
	Retarded resorption of the yolk sac
Production	
	Decreased egg size
	Goitre present in neonate
	Increased neonatal mortality
	Decreased productivity & longevity in breeders & layers
	Decreased egg production
	Decreased growth & weight gain
	Weight loss
	Decreased wool production & poor quality wool
	Increased body fat proportion
	Depressed milk production
General	
	'Mental sluggishness'
	Decreased food consumption
	Hair/wool dry & brittle or deficient
	Puffy appearance
	Decreased resistance to cold; preference for heat
	Decreased resistance (e.g., to infections)
	Stunted animals and dwarfed adults
	Nervous irritability
	Muscle tremors
	Long lacey feathers

it might be affected by climatic differences. For example, it was reported that 'hundreds of thousands' of hogs died annually in Montana from iodine deficiency before the addition of iodine to the hog diets (Mannar and Dunn, 1995) and one wonders whether cold weather contributed to these deaths and losses. A cold environment can certainly be lethal for a human patient with myxedema, and thyroidectomized chicks succumb to cold more quickly than intact animals (Nobukuni and Nishiyama, 1980), an effect which can be reversed by small amounts of thyroxine (Nobukni and Nishiyama, 1980). Perhaps these hogs might have survived, though not healthy, in a hot environment.

Livestock losses which probably cost millions of dollars annually may be caused by lack of iodine in animals diets, particularly in the developing countries. Losses in some of the less common animals have not been estimated, though information on some of these is becoming available now, such as the South American Camelids like Ilamas, alpacas, etc. (Pugh, 1995).

Selenium may be important for proper thyroid function, as it is an essential part of the deiodinase enzymes which convert thyroxine to triiodothyronine. Hence, some data on selenium requirements are also included here.

The occurrence of goitrogens, in rape, brassica, soybeans, etc. in the diet of cattle and other animals can raise the requirements for iodine, because of their interference with normal metabolism of that mineral.

As the above data indicate, the 'diagnosis,' considered on a worldwide basis, is incomplete. We can diagnose iodine deficiency in animals by the appearance of the above signs in the absence of deficiency of other nutrients, confirmed by very low tissue iodine content.

The geographic pattern of iodine deficiency in farm animals appears to follow the pattern of human iodine deficiency around the world (Pugh, 1995).

What about the treatment, or more precisely the prevention, of iodine deficiency in animals? There are numerous publications throughout the world giving tabulations of dietary requirements for various animals as well as some on the toxicity of substances added to fortify feed. Usually iodine is one of the nutrients included in the large lists, for example in textbooks at colleges of agriculture or veterinary schools.

Table 8.2 is a compilation of data from such tabulations.

Table 8.2. Estimated minimum dietary requirements of iodine

Animal	Body Weight (kg)	Iodine Requirement (microgram/day)
Poultry	2.3	5-9
Swine	68.2	80-160
Sheep	50.0	50-100
Dairy cow (in production)	454.5	400-800

Such gross aggregated data are of limited value to the farmer who must deal with animals at various times in their growth and development, or different phases of production. Here are two examples of tabulations giving iodine requirements for animals in various conditions (Tables 8.3 and 8.4)

Table 8.3. Dietary iodine, nutrients, and daily requirements of piglets & growing pigs

	Piglets		Growing Pigs		
Liveweight (kg)	1-5	5-10	10-20	20-60	60-100
Expected Weight gain (kg)	0.20	0.30	0.45	0.65	0.75
Expected feed intake (kg)	0.24	0.45	0.90	1.80	2.60
Expected efficiency (feed/gain)	1.20	1.50	2.00	2.75	3.45
Nutrient requirement (mg/kg diet) :					
Digestible energy (kcal/kg)	3,800	3,500	3,400	3,250	3,250
Iodine requirement	0.14	0.14	0.14	0.14	0.14
Selenium requirement	0.15	0.15	0.15	0.15	0.15
Amount per day (mg) :					
Digestible energy (kcal)	910	1,575	1,060	5,850	8,450
Iodine requirement	0.03	0.06	0.13	0.25	0.36
Selenium requirement	0.03	0.07	0.14	0.27	0.39

Table 8.4. Recommended levels of iodine (ppm) in feed for chickens, turkeys, and ducks

Chickens			
Starting Chicks (0-8 wk) --	Growing Chicks (8-22 wk) --	Laying Hens (22-40 wk) White Leghorn	Breeding Hens (22-40 wk) White Leghorn
0.35	0.35	0.30	0.30
Turkeys			
Starting poults (0-8 wk)	Growing turkeys (8-16 wk)	Finishing turkeys (16-24 wk)	Breeding turkeys --
0.35	0.30	0.30	0.30
Ducks			
Growing Ducks (0-7 wk)			Breeding Ducks --
0.30			0.30

These quantities for commonly eaten birds are lower than those given for mammals by Mannar and Dunn: they state that iodine levels for salt used for animals range from 30 to 400 ppm, certainly a wide range, and that if the iodine level in salt is at least 20 ppm it is likely to ensure that the minimum requirements will be met. They tabulate animal requirements as follows (table 8.5).

Table 8.5. Iodine requirements for animals*

Class of Animal	Iodine Requirement in Total Diet (ppm)	Salt Consump. (kg/yr)	Iodine content Required in Salt (ppm)
Swine	0.14	4.1	28
Beef Cattle	0.2-2.0	10	40-400
Dairy Cattle	0.25-0.5	24.3	50-100
Horses	0.1	10.9	20
Sheep	0.1-0.8	4.1	20-160
Goats	0.15-0.8	4-8	30-240
Poultry	0.3-0.4	0.2	120-160

* Table 17.1 from Mannar & Dunn, 1995 (in press).

The calculations become complicated, since one has to take into consideration not only the size of the animal but also how much feed it eats per day, as shown in tables 8.3 and 8.4 previously. It is impractical however, to titrate the amounts of added iodine to each level of growth and size, and generally one level is chosen and used for all.

Iodine is given to the animals in a number of different ways. The simplest is ground up meal of oyster shells and fish from the sea, which can provide adequate amounts of iodine when added to the grain fed to chickens and other edible birds. Chemical iodine may be added as salts of iodine, such as potassium or sodium iodide or iodate, and the organic iodine-containing compound EDDI is often used (ethylene diamine dihydr iodide). EDDI is recommended as a feed additive for livestock and poultry but not for dairy cattle in production.

Probably the most common method of providing supplemental iodine to mammals used by humans is via salt blocks, using iodized salt. For such animals in inaccessible regions, blocks of iodized salt can be dropped from airplanes.

Two other points should be mentioned.

First, the toxicity of iodine is relatively low, but it is not zero. The National Academy of Sciences of the US has published a series of books on animal nutrition (see endnote in references), and one of these is titled 'Mineral tolerance of domestic animals' which gives data on this point for iodine. Dietary levels of 200, 100, or even 50 ppm of iodine have been reported to retard growth and development. The animals develop signs similar to iodism in man; i.e., nasal discharge, coughing, etc. One author concluded that for calves levels higher than 25 ppm were undesirable and levels above 50 ppm were toxic. Other authors found no ill effects up to 40 ppm, but some adverse signs after 86 ppm (one study) or no effects even up to 283 ppm but adverse effects at 435 or above. The NAS book summarizes these various results. The range appears to be wide.

The results for laying hens are interesting. Egg production varied inversely with iodine intake from 625 to 5,000 ppm and ceased entirely above 5,000 ppm. These adverse effects in laying hens, however, were reversible. They resumed laying only one week after cessation of the high iodine diets.

The second and final point, one made by Mannar and Dunn, is the interchangeability of animal salt with human salt. In poor regions of the world, people may use for themselves salt intended for animals. This is an additional reason for making sure that so-called 'animal salt' is iodized, as well as human salt.

The authors thank Professor JAN Der-Fang, Ph.D., Faculty of Agriculture, National Chung-Hsing University, Taichung, Taiwan, for his invaluable aid, as well as Dr Huang Tien-Shang of National Taiwan University, Taipei.

References

Mannar, M.G.V., and J.T., Dunn. *Salt Iodization for the Elimination of Iodine Deficiency*. ICCIDD (International Council for Control of Iodine Deficiency Diseases), in press. Available through Professor John T. Dunn, M.D., University of Virginia School of Medicine, Charlottesville, VA 22908.

Mannar, V.M., and J.T., Dunn: *op cit.*

Nobukuni, K., and H., Nishiyama.1980. 'Effects of cold acclimation on survival of chicks in cold environment'. *Jpn. Poult. Sci.*, 17:91.

Nobukuni, K., and H., Nishiyama. 1979. 'Effect of a very small amount of thyroxine on survival of chicks exposed to cold'. *Jpn. J. Zootech. Sci.*, 50:676.

Pugh, D.G. 1995. 'Trace mineral nutrition in cattle'. *J. Bovine Practice.* In press.

Pugh, D.G. 1995. 'Trace mineral nutrition in South American Camelids'. *J. Vet. Med.* In press.

Pugh, D.G.: *op cit*

Endnote: These books may be obtained from the National Academy Press, National Academy of Sciences, 2101 Constitution Avenue NW, Washington, D.C., 20418, USA.

PART IV

Iodine Status of Feeds and Fodders

9

Iodine Status of Feeds and Fodders

S. P. S. Bedi

Abstract

The following paper represents a literature review on the topics of the iodine cycle in nature, manifestations of iodine deficiency in livestock, the role of iodine hormones, methods of estimation of iodine in feeds and fodders, the iodine status of different types of feeds and fodders, requirements of iodine in different animals, a discussion on various iodine supplements, and concludes with recommendations and a ten point action plan for the control of iodine deficiency disorders in livestock.

Introduction

India has a large human and livestock population estimated to exceed 1000 and 532 million, respectively, as well as 460 million poultry by the end of this century. There is a growing concern over the increasing population pressure, urbanization, and shrinking of cultivable land. The progressive decline in fodder crops and straw production due to the introduction of high grain yielding dwarf varieties of cereals and mechanical harvesting have further accentuated the gap between availability and demand for conventional feeds and fodders for livestock. Under these circumstances, animals are at great risk from the low availability of feed ingredients Furthermore, if the feeds are deficient in trace minerals (i.e., Cu, Fe, Mn, Co, I, Zn, etc.) the problems are compounded resulting in reproductive disorders, poor growth, and decline in milk production of other and animal products.

In recent years, relentless efforts have been made by the ICCIDD, WHO, UNICEF, and other agencies to eliminate iodine deficiency disorders affecting 1.6 billion people globally by the year 2000, but very little attention has been paid so far to determine the magnitude and the extent of damage caused by iodine deficiency disorders in livestock. Animals are an important part of the human food chain and contribute significantly to the agricultural framework and socio-economic conditions of the country. Therefore, their problems must be tackled judiciously along with the IDD elimination programmes for man.

It is taken for granted that wherever human IDD occurs, livestock may also be affected. Based on this assumption, about 100 million animals (cattle, buffalo, goat, and sheep) are in the IDD risk zones of the Himalayan and the sub-Himalayan regions of the country. Causing suffering to the animals and economic losses.

Iodine Cycle in Nature

The ecology of iodine deficiency in large areas of the world is due to the iodine cycle in nature. Iodine occurs in the soil and the sea as iodide. Iodide ions are oxidized by sunlight to elemental iodine which is volatile. The concentration of iodide in the sea water and air is about 50 μg/L and 0.7 μg/m^3, respectively. Iodine in the atmosphere is returned to the soil by rain, which has a concentration in the range of 1.8-8.5 μg/L. The return of iodine, however, is slow and small in amount compared to the original loss, and subsequent repeated flooding ensures that iodine deficiency in the soil continues. There is no natural correction and iodine deficiency persists in the soil indefinitely. All crops grown in such soils will be iodine deficient. As a result, human and animal populations, which are totally dependent on foods grown in such soil, become iodine deficient (Hetzel 1994).

Manifestation of Iodine Deficiency in Livestock

Goitre is a well known manifestation of iodine deficiency. It generally occurs in farm animals wherever human goitre is endemic and the factors that limit the capacity of the human thyroid to maintain structure and function apply equally to farm animals (Underwood, 1981). In the 1950s, the prevalence of goitre was reported in 10% of goats in Bareilly district as well as in hilly and sub-mountain areas of Uttar Pradesh, India (Ray *et al.* 1959). Goitre is often observed in newborn calves even when there are no clinical symptoms of iodine deficiencies in the dam (Hemken *et al.* 1971). Iodine deficient calves may be born blind, hairless, weak, or dead with the exact symptoms depending on the severity of the deficiency (Hemken, 1970). Likewise, iodine deficiency may reduce reproduction at any stage causing irregular or suppressed oestrus, abortions, still births and increased incidence of retained placenta. Male fertility may also be affected (Underwood, 1981). Dairy cattle fed insufficient iodine are less able to resist stress and may even have a higher incidence of ketosis (Hemken, 1970). Enlarged thyroid glands due to hypothyroidism and mild to moderate changes in the kidneys of goats fed *subabul (Sencaena Lencocephala)* has been observed (Sastry, 1989).

Role of Iodine Hormones

Iodine is an essential element which plays an important role as a constituent of thyroid hormones. The thyroid gland concentrates inorganic iodine from plasma and binds it to the amino acid tyrosine to form triiodothyronine (T_3) and thyroxine (T_4). Both are essential for normal growth, physical and mental development, maturation, and reproduction in both humans and animals alike. These hormones are stored by peptide linkage in thyroglobulin and released into the circulatory system by proteolysis. Approximately 0.5 per cent of T_4 and 0.5 per cent of T_3 in circulation is in the physiologically-active free form whereas the bound form acts as a reservoir. Elevated T_3 levels have been associated with thyrotoxicosis, toxic adenoma and toxic multinodular goitre, while decreased T_3 levels are found in surgical trauma, hepatic failure, renal failure and hypothyroidism. Elevated T_4 levels are associated with hyperthyroidism, hepatitis, pregnancy, and exogenous administration of oestrogen while decreased

T_4 levels are observed in hypothyroidism, chronic thyroiditis, and cretinism.

Thyroid insufficiency (impaired thyroid hormonogenesis) is not necessarily a reflection of inadequate supplies of dietary iodine. It can also be due to an inability or reduced capacity of the thyroid gland to accumulate iodine or to convert it into functional thyro-active organic combinations (Underwood, 1981). The determination of thyroid activities thus provides important information regarding thyroid function.

Estimation of Iodine in Feeds and Fodders

The first step in controlling IDD in livestock requires knowledge of the iodine content of various feeds and fodders in order to monitor its availability and intake by animals. Detailed information regarding the iodine content of different feeds and fodders during different seasons and in different climatic zones of the country in scanty.

Various chemical methods for the estimation iodine are:

1. Methods for the estimation of relatively larger amounts of iodine (e.g., iodized salts, mineral blocks, and mineral mixtures);

2. Methods for the estimation of small amounts of iodine (e.g., blood, milk, urine and water);

3. Methods for the estimation of very small amounts of iodine (e.g., feeds and plant materials).

The methods in the first category are comparatively simpler and with the current availability of spot testing kits, approximate concentration can be obtained in a given sample.

The estimation of iodine in blood, milk, urine, and water is complex, however, a simplified procedure has been developed, particularly for urine, providing a very useful tool in determining the present status of iodine intake given that about 90 per cent of ingested iodine is excreted through urine. This procedure, however, requires some sophisticated equipment which restricts its large scale application in the field.

Plants and feed material require the most difficult and complicated methods of estimation as they contain very small amounts of iodine (0.01 to 0.02 ppm). Methods generally involve fixation of plant iodine and oxidation and reduction processes requiring extra pure chemicals and well-equipped laboratory facilities.

Existing methods for the estimation of iodine in feeds and plant materials are quite tedious and time consuming. It is therefore, imperative to develop accurate, quick, and less cumbersome methods.

Iodine Status of Feeds and Fodders

Bedi (1993) reported the iodine content of different feeds and fodders (table 9.1).

These observations revealed that cereals, oil-cakes, and wheat bran are poor sources of iodine, while straws and green fodder contain low to marginally adequate levels of iodine as per the requirement of livestock.

Cereals, which form the bulk of daily feed intake of human beings, are poor sources of iodine. Observations made by Hetzel and Pandav, (1994), are substantiated with the above findings in that the large human population in the sub-Himalayan and terai areas are at risk of IDD due to the low ingestion of iodine. Furthermore, the iodine requirement of various categories of livestock may also be met as most of the ingredients of normal animal rations are either poor or marginally adequate sources of iodine. The daily intake of rations in most situations is inadequate in the necessary amount of iodine required for growth, reproduction, and production in animals.

Table 9.1. Iodine content of different feeds and fodders

Feed/Fodder	Iodine Content (μg/g DM)
Cereals (maize, wheat, barley)	0.115 - 0.192
Oil cakes (de-oiled GNC, mustard, mahua)	0.085 - 0.263
Brans, wheat, and rice (de-oiled)	0.095 - 0.287
Green fodders (maize, bajra, berseem, oat, subabul)	0.216 - 0.574
Straw (wheat, rice)	0.293 - 0.309

Source: *Bedi, S.P.S. 1993. Iodine status of livestock feeds and fodders. In: Nutrition Workshop Conference, Bhubaneswar, India.*

Factors influencing the availability of iodine and other trace elements

Availability of iodine and other trace elements required for livestock is influenced by various factors (i.e., type of soil, pH of the soil, content of the element in the soil, organic/inorganic fertilizers used, sources of irrigation, rainfall, crop species, crop rotation, stage of maturity, stage of cutting, climatic factors, interaction of the elements, as well as the age, sex, and species of the animals involved).

Underwood (1981) stated that the incidence of goitre is influenced by climatic and seasonal conditions through changes in pasture iodine content and productivity. Bedi (1991) observed large variations in the iodine content of berseem fodder from the month of December to April, i.e., in I, II, III, IV and V cut, each at one month intervals (table 2).

These results show that the iodine content declined with the fall in atmospheric temperature (i.e., around 10 °C in January, corresponding to IInd cut) but as soon as the temperature rose (in February and March), the iodine concentration improved (as in IIIrd and IVth cut). Further rising of atmospheric temperature, ripening, and flowering of the berseem, however, caused a decrease in the iodine content again (Vth cut).

The literature reveals that with an atmospheric temperature around 20 °C, root nodule activity increases the mobilization of nutrients (N, Ca, P, etc.) but that with a further rise in temperature (i.e., around 30 °C in April), the availability of nutrients falls. Similarly, it appears that atmospheric changes in temperature significantly influence the iodine content of berseem from the Ist to Vth cuts.

Earlier works of Bedi and co-workers (1994) show that livestock in the Himalayan and Sub-Himalayan regions suffer from trace element deficiencies (Cu, Fe, and Zn) as a result of their low abundance in the soil of the hills and foothills due to snow, rainfall, floods, and erosion.

The problem of trace element deficiencies is compounded by modern agricultural practices. High yields of crops are likely to lower the soil reserves of trace elements. In soils where the total amount of available nutrients or the absence of mother rock containing them is low to begin with, the deficiencies will be greater. The changing of crop patterns and the type of fertilizers used have been found to create imbalances of trace elements in soils, plants and animals.

In November 1993, the directors and scientists at the 13th Indian Council of Agricultural Research meeting at Lucknow, recommended that: 'Since iodine and phosphorus deficiency are causing great losses due to reproductive failures in our livestock, a survey about the occurrence of different diseases of nutritional and metabolic disorders may be carried out.'

Requirements of Iodine

Miller (1979) and Church (1979) reported that 0.25 ppm iodine in the feed dry matter will meet the minimum need of growing or non-lactating dairy cattle when no goitrogenic substances are eaten. Because of the secretion of iodine through milk, lactating cows require about 0.5 ppm (Miller 1979) or 0.6 ppm, (Church

Table 9.2. Iodine content of berseem fodder

Month (CUT)	December (I)	January (II)	February (III)	March (IV)	April (V)
Iodine content (µg/gDM)	0.196± 0.081	0.159± 0.078	0.351± 0.158	0.368± 0.106	0.303± 0.930

1979, Underwood, 1981) of iodine in their diet. Pregnant and high yielding cows require 0.8 ppm (ARC 1965). If the diet contains 25 per cent strongly goitrogenic feed, the iodine intake should be increased to at least double the requirements.

Bedi (1991) observed that *subabul*, a leguminous fast growing perennial green bushy fodder used for animal feeding contains adequate to high levels of iodine (0.3 to 0.5 µg/g DM). Shastry (1989) reported that feeding of *subabul* to female adult goats adversely affects fertility, neonatal mortality, and congenital goitre in the kids born. These studies also show that *subabul* fodder has a high content of mimosine (3-6%) permitting the inference that not only does mimosine inhibits the utilization of iodine, but that it also blocks the availability of iodine from other ingredients in the ration fed to animals. In such situations, the iodine requirements of animals are increased many fold.

Underwood (1981) reported that goitrogens are of equal or greater importance than low levels of iodine in feeds as contributing factors in iodine deficiency. Brassica species of plants, such as: kale, turnip, cabbage, as well as others like soyabean, linseed, peas, ground nut, and white clover contain high levels of goitrogens.

Iodine Supplements

In the case of iodine deficiency, the supplementation of potassium iodide, sodium iodide, or calcium iodide is recommended (Miller 1979). Dosing or drenching is effective in controlling the incidence of endemic goitre. Extra feeding of iodine compounds to prevent neonatal mortality and associated goitre in lambs over and above the requirements can be done with few doses (e.g., two doses of 280 mg KI or 360 mg KIO_3, one at the beginning of the fourth month and the other at the beginning of the fifth month of pregnancy).

Fig. 9.1. A spray-mix iodization unit operating in Tuticorin, India.

The most common and convenient supplement is iodized salt. The general recommendations for common salt supplement for livestock is 2 per cent of the concentrate or 0.5 per cent of the total dry matter intake. India has a large cattle, buffalo, goat, sheep, and pig population (i.e., 192, 70, 95, 40 and 10 million, respectively); the total population, including other animals, is well over 500 million. The daily iodized salt requirement for the average adult cow and buffalo is 40 - 50 g, 20 - 25 g for adult sheep, and 4.5 g (2 - 2.5 g for young) for goats and pigs. The total approximate annual requirement of iodised salt is about 5 million tonnes.

Recommendations

India produced 2.8 million tonnes of iodised salt for human consumption in the year 1992-93 (Hetzel and Pandav, 1994) as opposed to an installed capacity of 6 million tonnes. Thus, only about 50 per cent of the potential amount of iodised salt is being produced. Efforts need to be made to mobilize resources so that the salt processing capacity of the industry can be achieved. The enhanced capability of the industry should be utilized for the production of iodised salt for livestock. Animal grade iodised salt should be cheap and easily available to poor livestock owners in the remote areas of endemic regions. A 'ten point action plan for control of iodine deficiency disorders in livestock' is recommended for immediate consideration.

Conclusion

In general, iodine deficiency is endemic in Indian livestock which suffer from stunted growth and late maturity. Cows and buffaloes generally have a low productivity and young livestock are victims to poor growth and high mortality. The problems of iodine and other trace element deficiencies and imbalances in animals must be tackled through the constant monitoring of levels of trace elements in soils, plants, and animals and by adjusting animal rations as per the requirements (either with feeds rich in trace elements or with iodised salt supplements). Trace element studies in animals under controlled conditions are also required for different species under different physiological and climatic conditions.

Ten Point Action Plan For the Control of Iodine Deficiency Disorders in Livestock

1. Survey of the iodine status of livestock feeds and fodders in the different climatic zones of the country.

2. Establishment of iodine estimation facilities for water, soil, plants and animals in each district of the country.

3. Development and standardization of quick and precise methods for the determination of iodine in animal feeds and fodders.

4. Mobilization of resources for the production of about 5 million tonnes of iodised salt and making it available up to the village level.

5. Reduction in the cost of iodized salt for animal feeding.

6. Public education campaigns for the supplementation of iodized salt in animal rations.

7. Continuous monitoring of the effect of supplementation programmes on the health and production of animals.

8. Quick implementation of supplementation programmes with information from the village level passed on to district laboratories fortnightly and to state and zonal headquarters using computers.

9. Involvement and cooperation of the state animal husbandry departments so as to utilize the services of about 40,000 veterinary polyclinics, hospitals, dispensaries, and stockmen veterinary centres in the country.

10. Financial assistance from the government and international agencies.

References

ARC. 1965.*The Nutritional Requirements of Farm Livestock, No.2. Ruminants.* 'Agricultural Research Council: London'.

Bedi. S.P.S. 1991. 'Iodine status of berseem and oat fodder at different stages of maturity/cuttings'. In: 1st International Animal Nutrition Workshop Conference for Asia Pacific, Bangalore.

Bedi, S.P.S. 1993. 'Iodine status of livestock feeds and fodders'. In: Nutrition Workshop Conference, Bhubaneswar, India.

Church, D.C. 1979.'Digestive physiology and nutrition of ruminants'. O & B Books Ind.: USA p. 120.

Hemken, R.W. 1970. *J. Dairy Science.* 53: 1183.

Hemken, R.W., J.H., Vanderall, B.A., Sass, J.W., Hibbs. 1971. *J. Dairy Science.* 54: 85.

Hetzel, B.S., C.S., Pandav. (eds.). 1994. *S.O.S. For a Billion: the Conquest of Iodine Deficiency Disorders.* Oxford University Press: Delhi.

Miller, W.J. 1979. *Dairy Cattle Feeding & Nutrition.* Academic Press: New York.

Ray, S.N., B.N., Majumder, Braham Dutt, A.W., Zubary, P.K., Dwarkanath. 1959. *Final Report of the Scheme, Studies on the Influence of Trace Elements on the Health and Productive Capacity of Livestock in India.* I.V.R.I (ICAR): Izatnagar.

Shastry, M.S. 1989. *Annual Report.* I.V.R.I, Izatnagar. p. 32.

Underwood, E.J. 1981. *The Mineral Nutrition of Livestock.* Commonwealth Agricultural Burawx, England. p. 77-90.

10

Iodine Profile of Foods and Foliage

Rita Singh Raghuvanshi

Abstract

The iodine contents of water, soil, and foodstuffs in a goitre endemic area was estimated. The iodine content of water was determined by the method given by Dubrawcic (1955), while iodine content of soil and foodstuffs was estimated by the procedure given by Henry (1964). There was no difference in the iodine content of artesian well (0.0076 ppm) and hand pump (0.0064 ppm) water, whereas the drinking water in the endemic area contained less iodine than that from the non-endemic area (Hyderabad). Farm soil where cereals, pulses, and oilseeds were grown had a greater average iodine content (16.0 ppm) than kitchen garden soil (9.1 ppm) where mostly vegetables are grown. The iodine content of samples belonging to food groups such as cereals, pulses, fruits, leafy vegetables, roots and tubers, and other vegetables and flesh foods ranged from 0.10 to 13.33 μg/100g on a fresh weight basis. The iodine content of pulses was more than that of foods belonging to the other food groups. Wheat contained more iodine than rice. Meat was found to have an iodine content of 3.28 μg/100 g on a fresh weight basis.

Introduction

Iodine is a rate limiting substrate for thyroid hormone synthesis by the thyroid gland and must be available in threshold amounts to prevent goitre and hypothyroidism (Fisher, 1989). Thyroid hormones i.e. thyroxine (T_4) and triiodothyronine (T_3) function in regulating the basal metabolic rate and are essential in the growth and development of the brain and body. The human body has about 15 to 20 mg iodine of which about 70-80 per cent is present in the thyroid gland. Iodine deficiency results in the depletion of iodine stores in the thyroid gland which reduces daily production of T_4. A fall in the blood level of T_4 induces secretion of an increased amount of TSH (Kochupillai *et al.* 1984). When a deficiency of iodine persists for a long time, hyperplasia of the thyroid results in an effort to increase the efficiency of the thyroid gland to trap more iodine, manifesting itself in the form of goitre. Besides goitre, iodine deficiency disorders in humans include: abortions, stillbirths, increased infant mortality, cretinism, psychomotor defects, hypothyroidism, impaired mental functions, and retarded physical development (Hetzel and Dunn, 1989).

Iodine deficiency in animals has been observed in the form of reproductive failure and thyroid insufficiency. As a result of iodine deficiency, development of the foetus is retarded or arrested at some stage in gestation resulting in resorption, abortion or still birth, birth of a weak, hairless offspring associated with prolonged gestation and parturition, or retention of the placental membranes. In iodine deficient areas,

Table 10.1. Reported iodine content of selected foodstuffs

Reference and Notes	Foodstuff	Iodine Content	
		Endemic Area	Nonendemic Area
Mahadeva et al (1968)	Rice, country	4.3 - 21	88.00
	Bread	6.7	--
μg/100g	Wheat flour	8.7	--
wet weight	Green gram	2.2 - 28.1	88.00
	Leafy vegetables	0.8 - 39.0	45.00
	Brinjal	0.5 - 72.0	33.00
	Plantain	1.2 - 16.0	62.50
	Coconut	107.5 - 132.0	166.00
	Chili (green)	4.0 - 12.0	25.50
	Lime juice	(trace)	(trace)
	Curry Powder	4.4 - 84.7	98.00
	Sugar	300.00	--
	Tea	178.00	--
	Salt	8.50	--
	Water/L	4.4 - 4.6	182.00
Heinrich and Weak (1989)	Beet leaves and some brassicas	--	73 - 915
	Legumes and graminae	--	49 - 66
μg/kg DM	Field beans	--	209
	Sugarbeet tops	--	363
	Sugarbeet, fodder beet, and potatoes	--	74 - 91
	Water	--	0.5 - 6.7 μg/L

Table 10.1. (contd).

Reference and Notes	Foodstuff	Iodine Content Endemic Area	Non Endemic Area
Mageswaran and Sivasubramaniam (1984) - Marine algae mg/kg	Cracilaria optunica	–	1277.00
	G. Crassa	–	889.00
	Turbinaria ornata	–	810.00
NIN (1986-87) μg/100 g	Cereals and millets	–	13.6 - 73.0
	Pulses and legumes	–	10.0 - 56.0
	Leafy vegetables	–	2.7 - 20.0
	Other vegetables	–	1.4 - 4.6
	Roots and tubers	–	2.0 - 17.0
	Fruits	–	3.0 - 12.4
	Milk and milk products	–	36.0 - 86.0
	Fish	–	83.0 - 187.0

Source: *Mahedeva, Mageswaran and Sivasubramaniam, Heinrich and Wenk, NIN* supplementation of iodine has been reported to influence milk production in animals (Kobeisy *et al.* 1992), and egg production in poultry (Christensen, 1991). Excessive iodine intake has also been shown to increase the risk of goitre and thyroid carcinoma (Delange, 1988; Franchschi *et al.* 1990). Very few studies, however, are available in the literature (Mahadeva *et al.* 1968; Mageswaran and Sivasubramaniam, 1984; NIN, 1987; Heinrich and Wenk, 1989) which cite iodine content of foodstuffs (table 1). Since plants take iodine from soil and the animal's iodine requirement is met from plants (with a small amount from water), the present study has been undertaken to determine the iodine content of soil, water and foodstuffs in Pantnagar, Uttar Pradesh, India.

Methods and Materials

Water samples were collected from each type of drinking water source in seven blocks of the Pantnagar farm area. APHA12 as per labelled and stored under refrigeration.

Soils samples from farm land and kitchen gardens within the water sampling area were obtained from a depth of 15 cm, individually mixed, air dried, ground to pass a 2 mm sieve, and stored.

Food samples from various farm food groups, such as cereals, pulses, fruits, leafy vegetables, roots and tubers, and other vegetable and flesh foods were bought from the local market. Cereals and pulses were ground into powder form. Boneless and fatless meat was obtained for iodine estimation. All fruits and vegetables were washed, wiped, peeled (where required), chopped into small pieces and dried at 80 ^{0}C;

moisture content was subsequently estimated (AOAC, 1975). Dried samples were ground and stored in dry, air tight glass bottles. Estimations were made on triplicate samples using the reported procedures for water (Dubraucic, 1955), soil (Whitehead, 1973; Henry, 1964), and food samples (Henry, 1964).

In the preparation of a standard curve, 1 mL of each of the working potassium iodide standard solutions was dried at about 80 °C in 25 mL conical flasks. After complete drying, each standard sample was dissolved in 10 mL arsenous acid and the rest of the standardization procedure was the same as with the soil samples.

Results and Discussion

Iodine content of all sources of drinking water in the seven blocks of the Pantnagar farm is given in table 2. Iodine content of handpump water varied from 0.0040 to 0.0120 ppm with an average of 0.0064 ppm, while that of artesian well water ranged from 0.0053 to 0.0105 ppm with an average of 0.0076 ppm. However, no statistical difference in iodine content was found ($t=0.657$) between both water source types in Pantnagar. It is interesting to note that the iodine content of drinking water in Hyderabad (a non-endemic area) is greater than that of handpump water in Pantnagar (an endemic area) ($t = 11.333$), and artesian well water ($t = 11.273$).

The average iodine content of farm soil (16.0 ppm) was significantly higher than the 9.1 ppm found in kitchen garden soil ($t = 13.075$). Wheat, paddy, sugarcane, mustard and pulses were the main crops grown in the farm land while potato, onion, other vegetables, and at times paddy was grown in the

Table 10.2. Iodine content of drinking water in Pantnagar study area, Uttar Pradesh

Location	Iodine (ppm)	
	Hand Pump	Artesian Well
D Block	0.0066	0.0053
E Block	0.0066	--
F Block	--	0.0066
G Block	0.0040	--
I Block	0.0053	0.0079
J Block	0.0120	0.0105
K Block	0.0040	--
Average	0.0064±0.0030	0.0076±0.0022

kitchen gardens. This was in accordance with the results reported from Varanasi (Agarwal *et al.* 1988).

Iodine content of food samples belonging to various food groups is given on both dry and fresh weight basis in table 3. Iodine content ranged from 0.10 µg/100g in ridge gourd to 13.33 µg/100 g in black gram dal (C.D. at 5% level = 1.0954). Black gram dal contained more iodine as compared to other pulses. In general, pulses contained the highest iodine content (9.87 - 13.33 µg/100 g on fresh weight basis) followed by cereals (3.67 to 10.93 µg/100 g on fresh weight basis). Among cereals, wheat contained a higher amount of iodine as compared to rice. Among the four most commonly available fruits - apple, banana, guava, and papaya - apple showed the lowest amount of iodine (excluding papaya; C.D. at 5% level = 0.2105) as it comes from hilly regions where the soil contains less iodine. Guava and papaya are locally grown fruits while banana comes from the west and was observed to have the highest iodine content among fruits. Among vegetables, roots and tubers generally had a high iodine content (C.D. at 5% = 0.1030), however, no definite trend could be established among the different vegetables in the endemic area. In general, fruits and vegetables showed very low values of iodine. Meat was a good source of iodine with 3028 µg/100 g on a fresh weight basis. The comparable values of the iodine content of foodstuffs estimated in the present study was found to be lower than those values given by NIN (NIN, 1987), which estimated iodine the in foodstuffs from Hyderabad.

Table 10.3. Iodine content of foodstuffs from Pantnagar, Uttar Pradesh

Food Group	Foodstuff	Moisture (%)	Iodine (µg/100 g)	Standard Error of Mean
Cereals	Rice (Indrasan)	11.37	3.67	0.17
	Rice (IR 8)	12.45	4.40	0.40
	Wheat	7.16	10.93	1.07
Pulses	Black gram	10.90	13.33	1.33
	Lentil	7.03	12.00	0.00
	Red gram	8.21	9.87	1.07
Leafy vegetable	Cabbage	88.00	0.28	0.00
	Fenugreek	89.28	0.26	0.00
	Spinach	94.00	0.15	0.03
Other Vegetables	Beans	82.00	1.27	0.00
	Bitter gourd	92.84	0.13	0.00
	Bottle gourd	91.70	0.33	0.00
	Brinjal	89.75	0.49	0.04
	Cauliflower	88.10	0.25	0.04
	Cucumber	94.93	0.22	0.02
	Lady finger	88.64	0.20	0.00
	Peas	75.33	0.44	0.00
	Pumpkin	88.00	0.39	0.04
	Ridge gourd	94.21	0.10	0.00
Flesh foods	Meat	75.42	3.28	0.33

Table 10.3. (contd).

Food Group	Foodstuff	Moisture (%)	Iodine (µg/100 g)	Standard Error of Mean
Fruits	Apple	86.17	0.25	0.00
	Banana	69.67	0.99	0.11
	Guava	85.66	0.60	0.03
	Papaya	85.86	0.36	0.05
Roots and	Carrot	80.14	1.03	0.00
tubers	Colocasia	81.44	0.61	0.06
	Onion	86.11	0:45	0.05
	Potato	82.70	1.13	0.07
	Radish	92.91	0.54	0.05
C.D. at 5% level of significance				1.0954

Conclusion

Based on the results obtained and from perusal of the literature, it can be shown that the iodine content of foods grown in the goitre-endemic areas was in most instances, considerably lower than that in the non-endemic areas. No definite trend could be observed among the different vegetables sampled within the endemic areas. It was also found that the iodine content of farm and kitchen soils differs. Several factors might be responsible for both of these observations. Iodine in green fodder has been reported to be species related (Heinrich and Wenk, 1989; Groppel *et al.* 1989). Iodine content also decreases with the increasing of growth of plants (Heinrich and Wenk, 1989). It is possible that the degree of weathering and leaching of iodine in the sub-soil varies from place to place even in the same village (Mahadeva *et al.* 1968). As the iodine content of all foodstuffs used for human and animal consumption is low in Pantnagar, other dietary sources are needed. Iodised salt is one possibility, as well as marine algae rich in iodine for animal feeds (Mageswaran and Sivasubramaniam, 1984).

References

Agarwal, D.K., S., Srivastava, K.N., Agarwal, J.K., Agarwal. 1988. *Problems of Endemic Goitre in Gangetic Belt: Possible Measures to Control.* Nutrition Section, Department of Paediatrics, Institute of Medical Sciences, BHU, Varanasi.

AOAC. *1975. Official Methods of Analysis* (12th ed.). Association of Official Analytical Chemists: Washington, D.C. p. 394.

APHA. 1980. *Standard Methods for the Examination of Water and Waste Water (15th ed.)*. APHA - AWWA - W.P.C.F.: New York. p. 505.

Christensen, V.L. 1991. 'Iodine toxicity in large white turkey breeder hen'. *Poultry Science*. 70: 2402-10.

Delange, F. 1988. 'Role of goitrogenic factors other than iodine deficiency in the aetiology of goitre'. *Annalesd'Endocrinologie*. 49: 302-5.

Dubraucic, M. 1955. 'Determination of iodine in natural waters'. Analyst. 80: 295-7.

Fisher, D.A. 1989. 'Upper limit of iodine in infant formulas'. *J. of Nutrition*. 119: 1865-8.

Franchschi, S., R., Talamini, A., Fassina, E., Bidoli. 1990. 'Diet and epithelial cancer of thyroid gland'. *Tumori*. 76: 331-8.

Groppel, B., M., Ankic, B., Kulner, E., Scholz. 1990. 'Iodine deficiency in ruminants 1. The iodine content of feedstuffs, plants, and drinking water'. Archives of Animal Nutrition. 39: 211-20.

Heinrich, H., G., Wenk. 1989.'Iodine content of feeds and water used for cattle, sheep, and pigs in the Erfurt area'. 6th International Trace Elements Symposium. Vol. 3: 749-55.

Henry, R.J. 1964. *Clinical Chemistry: Principles and Techniques*. 'Hoeber Medical Division'. Harper and Row Publications. pp. 938-41.

Hetzel, B.S., J.T., Dunn. 1989. 'The iodine deficiency disorders: their nature and prevention'. *Annual Rev. Nutrition*. 9: 21-38.

Kobeisy, M.A., E.M., Ibrahim, M.M., Showtaewi. 1992. 'Composition and some technological properties of water buffalo milk as affected by dietary iodine supplementation'. *Assiut Veterinary Med. Journal*. 1992. 27: 111-22.

Kochupillai, N., M.M., Godbole, C.S., Pandav, M.G., Karmarkar, M.S.S., Ahuja. 1984. 'Neonatal thyroid status in iodine deficient environment of sub-Himlayan region'. *Indian J. of Med. Res.* 64: 537-44.

Mageswaran, R., S., Sivasubramaniam. 1984. 'Preliminary studies on the iodine content of some marine algae from coastal area of Jafna Peninsula'. *National Council of Sri Lanka.* 12: 173-78.

Mahadeva et al. 1968. *Brit. J. Nutrition.* 22: 527-34.

NIN. *Annual Report.* ICMR: Hyderabad. 1986-7. pp. 1-6.

Whitehead, D.C. 1973. 'Studies on iodine in British Soils'. *J. Soil Science.* 24: 262.

11

Feed Antimetabolites which Influence Thyroid Function in Livestock

T. Prasad and K.S.N. Prasad

Abstract

A literature survey on various feed antimetabolites which affect the iodine status of various livestock is presented. A discussion on environmental iodine deficiency caused by low levels of iodine in soil, water, feed, and fodder precedes the focus on elemental iodine antagonists and goitrogenic substances in foodstuffs which affect the uptake of iodine by the thyroid. Glucosides, mimosine, and nitrates, uracil, and thiourea are the principle chemical components found in various animal feeds and fodders that can cause manifestations of iodine deficiency disorders. The causative plant species and their effects on different livestock species is described.

Introduction

In many parts of the world, endemic goitre is found to occur in both humans and animals and is a recognized clinical condition of iodine deficiency. In animals , the major clinical manifestation is neonatal mortality with alopecia and visible or palpable enlargement of the thyroid gland in some animals. Enlargement of the thyroid gland is rare, but subclinical iodine deficiency - the death of a newborn without goitre - is of more importance. Iodine deficiency originates primarily from the following three factors:

i) Dietary iodine deficiency due to low levels in soil, water, feed and fodder;

ii) Presence of iodine antagonists, e.g., (a) Arsenic, F and Ca in excess amounts, (b) Cu and Mo in low amounts, (c) excessive N fertilization in forage production;

iii) Presence of certain goitrogenic factors (antimetabolites) in feeds and fodders.

In India, a systematic survey of iodine deficiency in livestock has yet to be conducted. Iodine deficiency origination from soil, water and feed, however, generally coincides with human goitrogenic regions, as in the Terai area including Assam, North Bihar, the hilly tracts of Uttar Pradesh, and some of the coastal districts of Maharashtra and Kerala.

Iodine Deficiency due to Soil, Water, Feed, and Fodder

The maximum iodine content of certain plants is controlled by strongly inherited factors and therefore, is independent of soil and season; perennial rye grass may contain 146 μg of I/100 g dry matter, whereas other

Fig.11.1. Worker feeds an iodization plant in a salt refinery near Dhaka in Bangladesh.

plants grown on the same soil may contain only 10 μg of I/100 g DM. Goitre in lambs is generally observed in those from ewes that consume less than 30 μg of I/100 g DM, a condition that can be precipitated by feeding white clover, subterranean clover, or rye grass. Logan *et al.* (1991) conducted a study of a dairy herd with a long history of the "syndrome" to monitor the effect of supplementing the feed of cows during the dry period with a commercial concentrate feed for one month before the expected date of calving. Initially, they had presumed that the condition was a manifestation of Se/Vitamin E deficiency. Group A received 1 kg of dairy concentrate, group B received the same quantity of dairy concentrate supplemented with 1 mg/kg selenium and 500 mg/kg vitamin E, and group C received no concentrate. There was no significant difference in the incidence of the 'syndrome' in groups A and B, however, both these groups showed a significant reduction in its incidence as compared to group C. They surmised that iodine deficiency might be a possible factor in the 'syndrome' as some of the affected calves demonstrated thyroid hyperplasia. Subsequently, they found that the incidence of death was markedly reduced after iodine was administered parentally to pregnant cows three months before calving.

In another study, Groppel *et al.* (1991) showed that the iodine content of a wide range of fodder plants, feeds, and of drinking water varied according to geological location. The distribution of white and red clover, green rye, and wheat (all at similar stages of growth) was observed to be influenced by soil iodine content. They also observed that concentrates of soybean oil meal, rape seed oil meal, and graminae were low in iodine. It was suggested that the iodine intake of ruminants via feeds and drinking water was influenced by geological location.

Dietary iodine deficiency was also observed by Salakhutdinov (1986) in a dairy farm complex where the iodine content was found to be 2.59 mg/kg DM for soil, 2.8 µg/L for drinking water, 155 µg/kg DM for green plants, and 171 µg/kg DM for feeds. The cows were clinically examined and histological findings indicated hypertrophy of thyroid glands suggesting goitre. The milk yield was also low. When the affected cows were each given three tablets containing 9 mg iodine, normal metabolism was re-established as well as normal levels of fertility and milk yield.

In India, Raina and Pachauri (1984) surveyed the region of Nainital and Rampur and observed that the prevalence of goitre in goats ranged from 0.16 - 6.00%. Iodine concentration of soils ranged from 5.5 mg/kg to 11.5 mg/kg, whereas water iodine content was normal. Postmortem examination of four young goats which died of iodine deficiency revealed hyperplastic epithelium and absence of colloidal matter in the acinar lumen of the thyroid gland.

Sewage water, if allowed to be consumed by animals, is likely to precipitate iodine deficiency. There are records of severe epidemics of goitrous calves from cows allowed to graze on pastures which were heavily dressed with crude sewage; prophylactic doses of potassium iodide given to grazing cows prevent such outbreaks.

Feed Antimetabolites and Iodine Metabolism

Feed antimetabolites affecting iodine metabolism can be categorized into three groups:

i) Goitrogenic glucosides

ii) Mimosine present in *subabul*

iii) Nitrate, uracil and thiourea

A number of cruciferous plants, e.g., rape seed (*Brassica napus*), mustard seed (*Brassica juncia*), and kale (*Crambea*) are valuable sources of protein. The protein content of these meals after oil extraction ranges from 30 - 50 per cent and amino acid composition is comparable to that of soybean. Although the protein quality of these plants in quite good, they are goitrogenic in nature. The goitrogens in these plants are glucosinolates which are capable of producing thiocyanate in the rumen.

Heavy intake of kale by pregnant ewes causes high incidence of goitre which is preventable by administering extra iodine in the diet. The thiocyanate production in the rumen varies with the variety of kale. Small young leaves produce five times as much thiocyanate as large fully formed leaves.

The high linseed meal diet of ewes contains linamerin and more than 20 per cent of this glucosinolate in the diet may result in goitrous offspring; supplementation of iodine inhibits this goitrogen. The glucosinolate present in mustard meal is known as sinigrin which when acted upon by the enzyme mirosinase produces ally1 isothiocyanate which is goitrogenic. The glucosinalate in rape seed is called progoitrin which when acted upon by the enzyme mirosinase produces goitrin.

Korber *et al.* (1985) surveyed cattle, sheep, and pigs for iodine deficiency diseases. 10 per cent of all livestock production units and 15 per cent of the pig, herds surveyed were found to be suffering from iodine deficiency. Inadequate iodine in the diet caused the disorder in 66 per cent of cases of sheep and in 20 per cent of cases of pigs and cattle. The remaining 80 per cent of cases of pigs and cattle were found to be related to the presence of iodine depressants (nitrates, thiocyanates, or glucosinolates) in the diet; clinical findings of prolonged pregnancies and goitres in the newborns confirmed this observation.

Most of these feeds cannot be avoided and so there is a need for greater supplementation of iodine to prevent iodine deficiency disorders.

Leucaena leucocephala (*subabul*), commonly referred to as Leucaena, contains 25 -35 per cent crude protein, but its use is limited by the occurrence of the toxic amino acid mimosine. Mimosine is converted to 3,4 dihydroxypyridine (DHP), in the rumen. DHP is a goitrogen impairing the incorporation of iodine into iodinated compounds in the thyroid gland, thus causing depressed serum thyroxine levels and goitre. Hamilton et al (1968) reported that four out of nine calves born to heifers fed leucaena had enlarged thyroid glands, three calves revealed signs of hyperactivity, and one calf with an enlarged thyroid was born dead.

Common crop and pasture plants and numerous weeds may accumulate toxic levels of nitrates. In ruminants, nitrate is readily reduced to nitrite which is absorbed and causes toxicosis. Abnormal accumulation of nitrate in plants is provoked by high levels of nitrogen fertilization. Chronic nitrate toxicity causes goitre, infertility, etc. Korber *et al.* (1986) observed that nitrate given to a batch of six cows (100 or 200 g/day for 28 days) reduced the serum protein bound iodine (PBI) and thyroxine levels as well as the iodine content in milk. On extra supplementation of 20 mg iodine/day, they observed a 31 per cent increase in serum iodine and a 15 fold increase in milk iodine. Similar effects were observed in parallel experiments with groups of ewes given nitrate at 12 or 24 g/day with or without iodine (5 mg/day).

Manifestations of Iodine Deficiency in Domestic Animals

In India, very little work has been done on iodine deficiency in domestic animals. Considerable work on this topic, however, has been reported abroad in the literature.

Gracia-Torregrosa (1992) reported that clinical iodine deficiency in a group of 12 breeding mares, diagnosed as abortion and goitre, was confirmed by analysis of milk and blood. Milk iodine values were below 20 µg/L for most of the year except in winter and early spring when the animals probably ate soil as they grazed. The T_3 and T_4 values in blood sera were generally below 0.7 µg/L and 15 µg/L, respectively.

Ghergariu *et al.* (1993) studied conditions of goitre calves in an area where goitre was endemic. The newborn calves on two farms had enlarged thyroid glands (some even died from asphyxia due to compression of the trachea by thyroid enlargement). Certain calves were also weak and had low body temperatures. On these farms, pregnant heifers and cows had been receiving two kg of soybean meal daily and this was considered to have a harmful effect. The serum protein bound iodine (PBI) values were very low in affected calves and their dam. Treatment of pregnant cows with 350 mg of potassium iodide weekly and of newborn calves with a hypodermic injection of an iodine compound improved clinical conditions of both groups.

In Orissa, Nayak (1988) screened 829 cattle during the period from 1971-88 for parasitic infestation of

Onchocerca. 108 animals manifested sclerotic lesions in the aorta due to Onchocerca infection. Out of these, 18 animals also manifested an enlarged thyroid gland. They concluded that the hypothyroidism was caused by iodine deficiency in the area as well as by infestation of the helminth based on histopathological evidence where sclerotic lesions were due to Onchocerca infection. Korber et al (1985) showed that endometritis and ovarian cysts were usually accompanied by impaired thyroid function.

Groppel *et al.* (1986) found that iodine deficiency affects normal growth in young sheep and goats and causes reproductive disorders in adults. In cows, iodine deficiency caused decreased milk production. Gurtler (1984) found that butanol-extractable iodine (BEI) and T_4 levels in blood of cows which gave birth to goitrous calves were significantly lower than those in cows giving birth to normal calves. The levels in the dam remained low up to 30 days after parturition. The BEI and T_4 levels in the blood of goitrous calves were also lower than in normal calves. The lower levels in goitrous calves could be replenished by intramuscular injection of 200 mg of KI between days 1-4 after parturition. Regression of goitre was accelerated by iodine injection.

Potter *et al.* (1985) administered iodized oil as a single intramuscular injection to pregnant iodine deficient ewes at 100 days of pregnancy and the subsequent growth of their foetuses was compared with that of fetuses of severely iodine deficient ewes and of iodine-sufficient ewes. All ewes were fed the same low iodine diet. The iodine supplemented revealed a remarkable improvement in thyroid function accompanied by an increase in brain growth. There was restoration of the number of cells (DNA) and myelination (cholesterol/DNA) in the cerebellum and cerebral hemispheres.

Conclusion

Thus, it is apparent that iodine deficiency in ruminants, and not simply clinical goitre, also influences the iodine status of the offspring. In addition to endemic environmental iodine deficiency, certain feed antimetabolites are also responsible for producing iodine deficiency disorders in livestock. Both conditions are controllable by supplying extra iodine in the feed as demonstrated in numerous studies.

References

Garcia-Torrengrosa, M.D. 1992. 'Iodine deficiency in horses'. *Nut. Abst. Rev.* (B) 62: 477.

Ghergariu, S., A.I., Baba, I., Papuc, G., Giurgiu, L., Kadar, V., Maries, R., Roca, E., Avaram, C., Orha. 1993. 'Simple non-toxic congenital goitre in calves in an endemic goitre region'. Revista Romana de medicina Veterinara. 3. (1): 36-54.

Groppel, B., R., Korber, M., Anke, A., Hening. 1986. 'Iodine deficiency in ruminants'. The Iodine content of feed stuffs, plants and drinking water. *Nut. Abst. and Rev.* (B) 61: 2478.

Gurtler, H., G., Pethese, R., Korber, P., Rudas, B., Turcke, R., Von Turcke. 1984. 'Influence of age of measure of thyroid gland function in calves with and without goitre and the influence of parenteral iodine administration'. Veterinary Bulletin. 54: 3776.

Hamilton, R.I., L.H., Donaldson, and L.J., Lambourne. 1968. *Aust. Vet.J.*44: 484.

Korber, R., N., Rossow, J., Otta. 1985. 'Iodine deficiency syndrome in cattle, sheep and swine'. Veterinary Bulletin. 55: 6457.

Korber, R., L., Jakel, H.J., Zastrow, H., Knoblauch, N., Rossow. 1985. 'Physiopathology of Iodine deficiency in cows in relation to their fertility'. *Nut. Abst. and Rev.* (B) 55: 3656.

Korber, R., B., Groppel, R., Leirer. 1986. 'Iodine and thyroid metabolism in cows and sheep given nitrate'. *Dai. Sci. Abst.* 48: 3874.

Logan, E.F., J.A., Smyth, D.G., Kennedy, D.A., Rice, W.A., Ellis. 1991. 'Stillbirth and perinatal weak calf syndrome'. *Veterinary Bulletin.* 61: 7949.

Nayak, B.C., L.N., Achoriyo, A.T., Rao, B.C., Das. 1988. 'Arteriosclerosis in cattle with concurrent thyroid cyst'. *Indian J. of Veterinary Pathology.* 12: 86-88.

Potter, B.J., M.T., Mano, G.B., Belling, D.M., Matin, B.G., Cragg, J., Chavadej, B.S., Hetzel. 1985. 'Restoration of brain growth in fetal sheep after iodized oil administration to pregnant iodine deficient ewes'. *Nut. Abst. and Rev.* (A) 55: 3043.

Raina, A.K., S.P., Pachauri. 1984. 'Studies on the prevalence of goitre in goats in Tarai'. *Indian Veterinary Journal.* 61: 8, 648-88.

Salakhutdinov, K.G. 1986. 'Iodine deficiency in farm animals'. *Nut. Abst. and Rev.* (B) 56: 6651.

PART V

Iodine Deficiency Disorders in India: Animal Studies

12

Iodine Deficiency Disorders in India: Animal Studies

M.G. Karmarkar

Abstract

In this paper, two historical experiments are described to illustrate the use of animal models (goats and rats) in studying the effects of iodine deficiency in humans and how the results can apply to livestock as well. The thyroid glands of goats from both iodine deficient (endemic) and iodine sufficient (non-endemic) areas were obtained after sacrifice and injected with radioiodine (^{131}I) to study its incorporation into iodoproteins for thyroid hormone synthesis. Evidence suggests that either iodine deficiency or a decrease in circulating TSH levels causes a decrease in T_4 hormone production which is essential for normal metabolism and a biochemical mechanism for this is proposed. To determine which factor is actually responsible for this drop in T_4 synthesis, a rat experimental model is described where four groups of rats were fed various supplemented diets to manipulate iodine uptake in the thyroid gland as measured again by injection of ^{131}I. Results support the hypothesis that iodine concentration is responsible for the differential aggregation of 27S and 19S iodoproteins which affects the level of T_4 synthesis. The use of animal models and the impact of iodine deficiency in both humans and animals is subsequently and briefly discussed.

Introduction

Iodine deficiency is amongst the most widely prevalent nutritional problems of the Indian peninsula. Goitre, the enlargement of the thyroid gland, is the major manifestation of iodine deficiency. Iodine deficiency has therefore, been historically equated with the prevalence of goitre in a community. The mountain slopes of the Himalayas, the Alps, the Pyrenees, and the Andes are known to have been the world's most notorious loci of endemic goitre (Kelly & Snedden, 1985). The northern frontiers of India, extending from Kashmir in the west to Assam in the east, form an extensive Himalayan goitre belt.

The pioneering work of Sir Robert McCarrison (McCarrison and Madhava, 1932) in the early part of this century focussed attention not only on the extent of the Himalayan endemic goitre, but also on its cause and on the reaction of the thyroid to various noxious influences (i.e., nutritional, toxic, and infective). Studies carried out by Stanbury and his co-workers (Stanbury *et al.* 1954) in Western Argentina and subsequently in other parts of the world (Holland, Finland, Greece, Eastern and Western New Guinea, and the Congo), established that environmental iodine deficiency is the major cause of goitre in these areas. There are a few reports (Clements, 1955; Tolley, 1951; Steyn, 1955) which incriminate other factors, especially goitrogens, in the etiology of endemic goitre. According to Hetzel (1970), however, environmental iodine deficiency is the main cause of endemic goitre with goitrogens playing an adjunct role in some areas.

In India, it is now established that apart from the Himalayan goitre belt, the problem of goitre is due to environmental iodine deficiency which exists all over the country in varying degrees of severity. According to the Ministry of Health, Government of India, out of 239 districts in 25 States and Union Territories which have been surveyed, 197 districts have iodine deficiency and endemic goitre. Thus, there is not a single state in our country which is free of iodine deficiency.

Research carried out in severe iodine deficient areas have shown that apart from goitre, several other problems exist, such as deaf mutism, spastic diplegia, stunted growth, mental retardation, and gait effects, all grouped under the syndrome of cretinism. Over the last two decades, studies have shown that all of the deformities encountered in the syndrome of cretinism are due to severe iodine deficiency during fetal and early developmental periods or are evident in later years of life (and so are irreversible in nature). Thus, the emphasis is now on the elimination of iodine deficiency rather than simply on endemic goitre. In order to avoid the irreversible damage of the deformities mentioned above in the future generations, priority should be given to the elimination of iodine deficiency amongst women in the reproductive age group and children in the developmental phase.

First Experiment – Goat Model

Introduction

Experimental models are extensively used in medical research. Iodine deficiency disorders have also been studied using experimental animals, like : rats, monkeys, marmosets, etc. Iodine deficient diets are usually given to these animals and include distilled water to produce the effect of iodine deficiency. The major drawback in such models is that it is difficult to mimic the same environmental conditions under experimental conditions; also, one has to arbitrarily choose the level of deficiency of a particular nutrient in the diet. In this experimental model, however, goats living in environmental iodine deficient areas were studied in an effort to preserve natural experimental conditions. An added advantage was that the thyroid glands from these animals could be obtained even in the smallest of villages as goats are usually sacrificed for meat purposes. It was first established that the thyroid glands of goats living in iodine deficient areas show similar pathophysiological changes as those found in humans living in the same area (exemplifying the usefulness of animal models).

Methods

Goats from endemic areas were given 75 to 100 μCi of ^{131}I, 48 hours prior to being killed. The thyroid glands were collected in ice at the time of sacrifice and brought to the laboratory for analysis while maintaining cold conditions. Thyroid glands from Delhi goats were also collected in a similar manner and served as controls. Incorporation of ^{131}I in iodoproteins of these glands was studied using polyacrylamide gel electrophoresis and the incorporation in iodoaminoacids was estimated by paper chromatography (Karmarkar *et al.* 1969).

Results

From table 1, it is clear that the thyroid glands of goats from the endemic area have higher MIT/DIT and T_3/T_4 ratios than those from the Delhi (control) area. Table 2 shows the incorporation of ^{131}I in iodoproteins of the thyroid glands of goats from endemic and Delhi area. There was a greater incorporation of ^{131}I in 27S iodoproteins in glands from the endemic area than in glands from Delhi. Consequently, less ^{131}I incorporation in 19S iodoproteins was observed in glands from the endemic area as compared to glands from Delhi.

Table 12.1. Distribution of iodoamino acids in the goat thyroid glands from endemic and control areas

Study Area	Origin	DIT	MIT	I	T_4	T_3
Endemic (n=10)	0.70± 0.16	35.90± 1.71	39.10± 2.42	16.30± 2.42	2.40± 0.43	6.20± 0.75
Control (n=10)	3.30± 0.35	53.80± 1.26	30.80± 0.89	6.40± 0.98	4.30± 0.47	0.90± 0.10

Table 12.2. Gel electrophoretic pattern of goat thyroid iodoproteins from endemic and control area

Study Area	33S	27S	19S	12S	8S	3S
Endemic (n=7)	6.40± 1.87	24.70± 4.34	65.80± 3.41	1.10± 0.39	1.20± 0.24	0.90± 0.29
Control (n=9)	5.50± 0.33	10.90± 0.33	80.30± 0.75	1.50± 0.27	1.00± 0.23	1.00± 0.17

Discussion

The results of the analysis of goat thyroid glands from the endemic area indicate that these glands are synthesizing more T_3 and T_4 as compared to goat glands from Delhi. The mechanism of thyroid hormone synthesis has been investigated in detail by Taurog and Howells (1966). They showed that the tertiary structure of thyroglobulin (19S iodoprotein) is important in the coupling reaction of iodotyrosine to form thyroid hormones. The coupling reaction is intramolecular and is not enzyme-mediated (De Crombrugghe *et al.* 1967). The tertiary structure of thyroglobulin is such that the DIT residues are situated in a favourable position at the coupling site to form T_4. In iodine deficient glands, however, there are more MIT than DIT residues and it is probable that at the coupling site in thyroglobulin, instead of two DIT residues, there maybe one MIT and one DIT residue which would result in the production of more T_3. Thus, in the iodine deficient gland, there is more production of T_3 than T_4 which may be viewed as a consequence of iodine deficiency.

The iodoprotein distribution in iodine deficient glands showed a higher proportion of 27S and less 19S (thyroglobulin) as compared to goat thyroid glands from the Delhi area. The 27S iodoprotein has the same subunits as that of 19S iodoprotein but has a higher content of iodine and thyroxine, as compared to 19S iodoprotein on a molar basis (Salvatore *et al.* 1965). Thus, in the thyroid glands from the endemic areas, the increased synthesis of 27S iodoprotein, which is more iodinated and can produce more T_4 hormone mole per mole than 19S, can be viewed as an adaptive mechanism of the gland in response to iodine deficiency. It is logical to consider that just as 19S iodoprotein under iodine deficiency can produce more T_3 than T_4, 27S iodoprotein may also do so under similar conditions provided that the mechanism of the coupling reaction is similar for both iodoproteins.

Conclusion

The difference between the iodine deficienty glands and control glands is due to the influence of circulating TSH and iodine concentration in the gland. In order to determine which one of these is responsible for the changes observed in the goat thyroid gland, an experimental rat model was subsequently employed in a second experiment.

Second Experiment — Rat Model

Introduction

Two drugs, perchlorate and propylthiouracil (PTU), can be added to the diet of albino rats to induce different concentrations of iodine in the thyroid gland. Perchlorate blocks iodide absorption in the gland while PTU blocks the coupling reaction. Stock diets with either of these drugs or a low iodine content raise the concentration of TSH in the circulation. This experimental design yields different influences of TSH and different concentrations of iodide in the thyroid gland which can be used to provide an answer to the query raised in the conclusion above.

Methods

Forty albino rats weighing between 130g and 160g were divided into 4 groups of 10 rats. Group A received a stock diet with an adequate amount of iodine; group B was given a low iodine diet obtained from Nutritional Biochemicals Corporation, Cleveland, Ohio; group C received a stock diet with an adequate amount iodine plusan oral dose of 150 mg of potassium perchlorate/day; group D received a stock diet with an adequate amount of iodine plus an oral dose of 5 mg of PTU/day. The rats were kept on the above diet schedules for 45 days after which each rat was injected with 5μg of ^{131}I intraperitoneally 24 hours before being killed. The thyroid glands were removed and ^{131}I incorporation in iodoproteins of the glands was determined.

Table 12.3. Gel electrophoretic pattern of rat thyroid proteins in different experimental groups

Group	No. of Rats	33S	27S	19S	12S	8S	3S	Total Iodine in Gland (μg)	Total Weight of Gland (mg)
Stock Diet	10	2.10± 0.24	10.04± 0.65	83.00± 0.86	2.70± 0.42	1.22± 0.18	0.72± 0.13	93.00± 3.78	28.00± 1.96
Low Iodine Diet	10	4.58± 0.46	26.35± 0.50	66.06± 0.87	1.45± 0.36	0.60± 0.11	0.51± 0.09	24.00± 1.60	54.50± 4.37
$KClO_3$ Treated	10	5.00± 0.34	30.00± 0.85	61.40± 0.74	2.15± 0.23	1.21± 0.25	0.61± 0.10	15.50± 0.88	67.50± 3.11
PTU Treated	10	2.47± 0.28	19.35± 0.62	74.20± 1.03	2.11± 0.32	1.11± 0.19	0.68± 0.14	25.00± 2.71	66.00± 2.33

Results

^{131}I incorporation in 27S iodoproteins was highest in the group receiving potassium perchlorate followed by the low iodine group, the PTU- treated group, and finally the stock diet groups (table 3). Consequently, and in an inverse manner with 27S iodoproteins, there was a reduction of ^{131}I incorporation in 19S iodoproteins in these groups.

Discussion

^{131}I incorporation in 27S iodoproteins seems to be universally related to iodine concentration in the gland. The results obtained using rats indicate that iodine concentration has more influence on the differential aggregation of the subunits to form either 27S or 19S iodoproteins than to form TSH.

Conclusion

Animal models have been used by other investigators to demonstrate the effect of thyroxine (T_4) on the brain and central nervous system during fetal and early development at periods. All of these studies have been conducted in order to understand the role of iodine in the consequences of its deficiency, especially the decrease in T_4 observed, so as to take the necessary corrective measures in humans. It is essential, however, to recognize that environmental iodine deficiency also has similar deleterious effects on livestock as in humans. In view of the importance of livestock in the economy, it is necessary to tackle the situation of iodine deficiency in livestock with the same determination as with humans. Research between investigators in the medical and veterinary sciences is currently being united in an effort to collaborate in efforts to eliminate iodine deficiency disorders in both man and livestock given the intimate interrelationship in the health of both.

References

Adaptation of thyroid gland to iodine deficiency. Life Sci. 8: 1135.

Clements, F.W. 1995. 'A thyroid blocking agent as a cause of endemic goitre in Tasmania'. Preliminary communication. *Med. J. Australia* 2: 369.

De Crombrugghe, B., H., Edelhoch, C., Beekers, M., De Visscher. 1967. 'Thyroglobulin from human goitres'. *J. Biol. Chem.* 242: 5681.

Hetzel, B.S. 1970. 'The control of iodine deficiency'. *Med. J. Australia* 2: 615.

Karmarker, M.G., N., Kochupillai, M.G., Deo, V., Ramalingaswami. 1969. 'Adaptation of thyroid gland to iodine deficiency'. *Life Sci.* 8: 1135.

Kelly, F.C., W.W., Snedden. 1958. 'Prevalence and geographical distribution of endemic goitre'. Bull. World Health Organ. 18: 5.

McCarrison, R., R.B., Madhava. 1932. 'The life line of thyroid gland'. *Indian J. Med. Res.* Memoir No. 23. Calcutta.

Ramalingaswami, V. 1953. 'The problem of goitre prevention in India'. Bull. World Health Organ. 9: 275.

Salvatore, G., G., Vecchio, M., Salvatore, H.J., Cahumann, J., Robbins,1965. '27S Thyroid iodoprotein: isolation and properties'. *J. Biol. Chem.* 240: 2935.

Stanbury, J.B., G.L., Brownell, D.S., Riggs, J.Itioz., Perrivetti, B., del Castilloe. *Endemic Goitre: the Adaptation of Man toIodine Deficiency*. Cambridge, Mass. 1954.

Steyn, D.G. *Endemic goitre in Union of South Africa and Some Neighbouring Territories*. Pretoria. 1955.

Taurog, A., E.M., Howells. 1966. 'Enzymatic iodination for tyrosine and thyroglobulin with chloroperoxidase'. *J. Biol. Chem.* 242: 1329.

Tolley, M.A.T. *Annual Report of Director-General of Health, New Zealand*, 1950-1951. Wellington, New Zealand. 1951.

PART VI

Control Measures for IDD in Livestock

13

Control Measures for Iodine Deficiency in Livestock: Current Status

R.S. Sachan and T.R. Rathore

Abstract

As in human beings, the deficiency of iodine in livestock is widespread, particularly in the mountainous and high rainfall areas of the world, expressing itself through many symptoms and disorders. This paper reviews the status of available control measures and their respective suitability under a variety of situations. Various iodine compounds and natural products in the form of fortified salt, vegetable oil, aqueous solutions, capsules, drops, meals, etc. are available for various situations of prophylaxis or therapy and scales of operation.

Introduction

Iodine deficiency is recognized by public health authorities as the most common preventable cause of metal defects and goitre in the world today. The World Health Organization (WHO) has estimated that correction of iodine deficiency would prevent brain damage that causes irreversible mental handicap in at least 20 million people in the world. The same problem in animals also, calls for remedial action.

Iodine was abundant during the primordial development of the earth, but large amounts were lost from the surface soil because of glaciation, snow and rain, as well as river erosion which carried it to the sea. Most iodine exists in the oceans at a concentration of about 50 μg/L. It also occurs in the deeper layers of the earth and is found in oil wells and natural gas effluents. Iodine occurs in soil and seawater mainly as iodide. Iodide ions are oxidized by sunlight to elemental iodine which is volatile and according to one estimate, some 400,000 tonnes escape every year from the surface of the seas to the atmosphere where the concentration is about 0.7 $\mu g/m^3$. Iodine in the atmosphere is returned to the soil through rain in concentrations of 1.8 - 8.5 μg/L (Hetzel, 1994). However, this is not sufficient replenishment and so an imbalance in soil is created.

In general, the older and more exposed the soil surface, the more likely it is to be leached of iodine. Thus, the most likely areas to be deficient in iodine are the high rainfall and mountainous areas of the world, such as the Himalayas, the Andes, the Alps, and the vast mountains of China. Environmental iodine deficiency is also known to occur in flood river valleys, such as the Ganges in India (Pandav & Kochupillai, 1984).

An indication of soil iodine content is conveniently obtained from iodine levels in the local drinking water. As compared to an average level of 9.0 μg/L of iodine in the city of Delhi, which is rated as only mildly iodine deficient, parts of Nepal approach 2 μg/L and in the severely iodine deficient areas of India, 0.1 - 1.2 μg/L of iodine is not uncommon. Although manifestations of Iodine Deficiency Disorders (IDD) have been recognized in human beings since ancient times, IDD in animals has only recently received attention. Geographical locations where iodine deficiency is more pronounced in human beings are likely to be the same as for animals. In addition, the problem in animals may be more severe due to a relatively closed foodchain, whereas humans also consume a variety of foods imported from non-deficient areas.

The role of iodine in the animals body is solely as a constituent of thyroxine and other related compounds synthesized by the thyroid gland. A deficiency of iodine therefore, causes structural and then physiological abnormalities of the thyroid gland which affects the entire animal body. The adult animal body contains less than 0.6 mg/kg of iodine, one third of which is concentrated in the thyroid gland, the remaining in other tissues. The ovaries, muscles, and blood, tend to have the highest concentrations of iodine next to the thyroids. In calves, the average iodine content of a particular organ varies over a narrow range: whole blood (0.063 - 0.078 ppm), blood serum (0.061 - 0.074 ppm), hair (0.017 - 0.032 ppm), hide (0.027 - 0.040 ppm), liver (0.080 - 0.081 ppm), kidneys (0.070 - 0.072 ppm), lungs (0.050 - 0.052 ppm), and thyroid gland (240 -380 ppm).

Iodine is absorbed as iodide from any portion of the alimentary tract, most readily from the small intestine. On reaching the thyroid gland, iodide is quickly oxidized to iodine and bound to tyrosine molecules of the thyroglobulin, a glycoprotein. In the thyroglobulin molecule, iodine is present primarily as 3-monoiodotyrosine, 3,5-diiodotyrosine, 3,5,3-triiodothyronine, and 3,5,3,5-tetraiodothyronine. On the hydrolysis of thyroglobulin, thyroxine and other iodinated compounds are released into the bloodstream. In areas of iodine deficiency, development of the animal fetus is retarded or arrested at some stage during gestation resulting in early death or resorption, abortion or still birth of weak, hairless offspring associated with prolonged gestation, or parturition and the retention of the placental membranes. Subnormal thyroid hormone levels in herds of cattle are accompanied by a high incidence of aborted, stillborn, and weak calves (Hetzel, 1989).

In order to prevent and correct iodine deficiency in animals, various means have been followed and advocated in different parts of the world. The choice of control measures in animals largely depends on the severity of the problem in degree and numbers affected and on the economic stakes involved. Each of the different control measures available for the prophylaxis and therapy of iodine deficient livestock is discussed below.

1. Iodized salt

Two forms of iodine are used to iodize salt: iodide, as in 'iodized salt' and iodate, as in 'iodated salt'; both, however, are generally referred to as 'iodized salt' in common usage. Iodate is less soluble and more stable than iodide and is therefore, preferred for humid tropical conditions. The mixing of iodine with salt is simple and without any adverse chemical reaction. The processes used for mixing are: dry mixing, drip feeding, and spray mixing.

Since iodine is required in very minute quantities (150 - 200 μg/day), the dosage of iodine in salt is extremely low. The normal iodine concentration in salt for human consumption is in the range of 30 -100 micrograms of iodine per gram of salt. Such iodized salt can be easily administered to animals either orally or mixed with animal feeds, fodder, and concentrated for regular prophylactic use in endemic areas or as specific therapy when required. Similarly, suitable salt licks can also be prepared and placed at appropriate locations in animal yards.

Several studies address the rate at which iodine should be administered to cattle. It is generally recommended that the iodization level of salt for animal consumption should be 200 ppm (200 mg/kg) using potassium iodate. Blood *et al.* (1979) suggested, however, that potassium iodide given alone is unsuitable, but when mixed with calcium stearate (8% stearate in potassium iodide), it can be added to salt at the rate of 200 mg per kg of salt. Piel (1979) reported that oral treatment of goitrous calves with 2.5 mg of kg daily for 8 days or provision of iodized salt licks produced a satisfactory restoration of normal thyroxine levels. Anon (1959) reported that endemic goitre can be prevented by the addition of iodized salt to the diet with a daily intake of iodide not to exceed 200 - 300 μg.

2. Iodized water

The other convenient way of providing iodine to livestock is through drinking water. The appropriate amounts of potassium iodide or potassium iodate can be added to water containers, drinking water wells or water pipelines. Such treatments would be required only occasionally depending on the severity of the problem.

In some countries, chemical iodine is used to purify water and reduce bacterial pollution, which certainly benefits iodine deficient human and animal populations in the area. Although water iodization has been tried in the range of 50 to 500 μg/L, a dose of less than 120 μg/L has been found to be ideal from the point of view of odour (Dunn, 1994).

The use of porous polymer vessels containing iodine in community water supplies has been reported by Fisch *et al.* (1993) and could be conveniently utilized for animal systems with appropriate adjustments made in dosage.

3. Direct iodine supplementation

Iodine in the form of tablets or drops has been used in some countries to correct iodine deficiency in man. Tablets containing 100 - 500 μg of potassium iodide are available for daily consumption (as in Germany). Direct use of oral potassium iodate in pregnant ewes has been reported by Rodel (1971) and in lambs by McCauley *et al.* (1973). Direct use of potassium iodide in poultry feed at the rate of 500 ppm has been reported by Roland *et al.* (1977) for IDD therapy.

Mason and lady (1978) reported that pregnant ewes in a goitre -prone area of Northern Tasmania were treated with an intraruminal device designed to release a regular quantity of iodine into the rumen.

Lugol's solution (potassium iodide) containing 6 mg of iodide per drop has been widely used as a disinfectant for wounds and for oral prophylaxis of iodine deficiency in human beings. Such supplementation could be adapted for the correction of iodine deficiency in livestock. Rajkumar (1970) reported that Lugol's solution in drinking water, oral administration of potassium iodide to pregnant goats, and intramuscular injections of an iodine preparation in affected kids were found to be effective prophylactic treatments.

George (1966) observed that the incidence of goitre in lambs born in spring and summer was reduced when ewes were given 0.4 g iodine intramuscularly as an organic iodine preparation approximately one month before lambing. This same treatment, however, had no effect on the survival of lambs born in winter, spring, or summer. Seidel *et al.* (1977) found that intramuscular treatment with potassium iodide in aqueous was form effective and resulted in an increase in protein bound iodine (PBI), weight gain, and reduction of thyroid size in cattle.

The normal recommended dietary intake of iodine for cattle is considered to be 0.1 - 0.3 mg/kg dry matter of feed.

4. Iodine-rich natural products

Certain seaweeds, such as Sargasso and *Laminaria japonica tresch* and thyroid glands of animals have been known to be effective in controlling IDD since ancient times. In more recent times, powders and extracts of these iodine-rich sources have been used for medicinal purposes. These natural sources could be used to fortify the animal diet. The use of seaweed meal containing 1 to 10 per cent iodine has been reported by Romijin (1955) in poultry production.

5. Application of iodine to soil-plant systems

It has been established that soils of mountainous regions and other eroded and leached areas have a negative iodine balance. In order to improve the iodine status of such areas of the biosphere, it may be prudent in certain situations to apply iodine-rich chemicals to the soil-plant system. Pastures and fodder-producing areas may be targeted for this purpose. The safe range of application would depend upon the sensitivity of various plant species to the addition of iodine. Russian scientists have tried the use of seaweeds as fertilizers (Beaton and Bengoa). Careful investigation is necessary before a rational soil application technology can emerge.

There have been several attempts to determine safe levels of iodine applications to the soil where phytotoxicity is noted (Sachan, 1994). As iodine is not essential for plant growth or forage yields, soil application would raise the tissue concentration of this element and subsequently act as a vehicle for administering iodine to animals through the foodchain which would eventually lead to positive effects for the entire ecosystem, including humans.

Iodine Toxicity

There have been instances where administration of iodine to human and animal systems has proved toxic. Therefore, adequate caution should be exercised when taking up therapy or IDD elimination programmes in livestock. The concentration of iodine in various carriers, dose, and frequency of administration should be carefully standardized depending upon the targeted animal species and the scale of operation.

McCauley *et al.* (1973) attempted potassium iodide doses in lambs ranging from 94 to 785 mg per day per lamb. They found that lambs treated with higher doses were more lethargic, consumed less feed, and gained less weight than lambs given smaller doses. Blood *et al.* (1979) also reported harmful effects of higher doses of iodine in animals. The main symptoms in cattle were: anorexia, severe pityriasis, restlessness, hyperthermia, and bronchopneumonia. Newton (1973) reported the development of symptoms of toxicity in pigs at 400 and 800 ppm levels of potassium iodate. For calves, the minimum toxic level has been identified as 50 ppm; even 25 ppm was found to be undesirable (Newton, 1974). An early study by Malan *et al.* (1932) reported the detrimental effects of iodine on reproduction when 0.02 g of potassium iodide was given as a daily supplement to Merino ewes for 16 months.

Conclusion

As with humans, many iodine-containing compounds (elemental iodine, iodides, iodates, organic iodine preparations, seaweed meal, thyroid gland meal, etc.) have been used in livestock to prevent and treat iodine deficiency. Vehicles of iodine include: common salt, vegetable oils, water, feed, soil, plants, and even some specifically designed porous vessels and intraruminal devices. The modes of administration include oral (water, capsules, drops) and intramuscular oil injections, directly or with different carriers. The choice of method may depend on the objectives (prophylaxis or therapy), targeted livestock species, scale of operation, and economic considerations.

References

Anon. 1959. 'Endemic goitre'. *Brit. Med. J.* 3: 36-7.

Beaton and Bengra.

Blood, D.C., J.A., Henderson, O.M., Rodostits. 'Veterinary Medicine'. 5th ed. *Bailliere Tindall: London.* 1979.

Dunn, J.T. 1994. 'The use of iodized oil and other alternatives for the elimination of Iodine Deficiency Disorders'. In: Hetzel, B.S., C.S., Pandav, (ed's.). *S.O.S. for a Billion: The Conquest of Iodine Deficiency Disorders.* Oxford University Press: New Delhi. 1994. Chap. 6.

Fisch, A., E., Pichard, T., Prazuek, et al. 1993. 'A new approach to combating iodine deficiency in developing countries: the controlled release of iodine in water by a silicone elastomer'. *Amer. J. Public Health.* 83: 540-5.

George, J.M. 1966. 'The use of an organic iodine preparation for controlling goitre in lambs born in winter, spring and summer'. *Proc. Aust. Soc. Anim. Prod.* 6: 403-4. (Cited in Vet. Bull. 37: 655).

Hetzel, B.S. 1989. 'The Story of Iodine Deficiency: An International Challenge in Nutrition'. Oxford University Press: Delhi.

Hetzel, B.S., C.S., Pandav. (eds.) *S.O.S for a Billion: the Conquest of Iodine Deficiency Disorders*. Oxford University Press: Delhi. 1994.

Malan, A.T., P.J., Dutoit, J.W., Groenewald. 1932. 'Studies in mineral metabolism XX, iodine in the nutrition of sheep'. 18th Rep. Direct. Vet. Serv. and Anim. Indust. Union of South Africa, Part II, pp. 651-76. (Cited in the Vet Bull. 3: 724-725).

Manson, R.W., R., Lady. 1978. 'Prevention of ovine congenital goitre using iodine - releasing intraluminal devices: preliminary results'. *Australian Journal of Experimental Agriculture and Animal Husbandry*. 18: 653-57.

McCauley, E.H., L.G., Linn, R.D., Goodrich. 1973. 'Experimentally induced iodine toxicosis in lambs'. *Amer. J. Vet. Ras*. 34: 65-70

Newton, G.L. 1974. *J. Anim. Sci*. 38: 499.

Newton, G.L., A.J., Clawson. 1974. *Anim. Sci*. 39:879.

Pachauri, S.P. 1980. 'Blood cellular and biochemical studies on goitre in goats'. *Indian J. Vet. Med. Association* (in press).

Piel, H.P. 1979. 'Goitre in the calf: detection and treatment of iodine deficiency in herd of cattle in combrailles, France'. *Recucil de Medicine Veterinaire*. 155: 605-10. (Cited in the in the Vet. Bull. 50: 6320).

Rajkumar, S.S. 1970. ' Incidence of goitre in goats'. *Ind. Vet. J*. 47: 185-7.

Rodel, M.G.W. 1971. 'Neonatal goitre and skeletal deformities in autumn born lambs on star grass pasture'. *Rodesia Agricultural Journal*. 68: 109-10. (Cited in the Vet. Bull. 42: 2883).

Roland, D.A. Sr., S.T., McCready, R.H., Stonerock, R.H., Harms.1977. 'Hypercalcemic effect of potassium iodide on serum calcium in domestic fowl'. *Poultry Science*. 56: 1310-14.

Romijin, C. 1955. 'Iodine and thyroid activity in the fowl'. *Acta Physiol. Pharm. Ne'erl*. 4: 121-2. (Cited in Vet. Bull. 26: 3566).

Sachan, R.S. 1994. 'Iodine in Soils'. Paper presented at IDD and Child and Women Development Workshop held at G.B.P.U.A. & T., Pantnagar, 19-20 January, 1994. Organized by NSS/UNICEF.

Scidel, H., J., Klahn, H., Kiupel, G., Blodow, P., Heller, H., Loapelmann. 1977. 'Iodine deficiency syndrome of the fattening'. *Zid Monatshefte fun Veterinarmedizin*. 32: 693-7.

Thilly, C.H., F., Delange, L., Ramioul, R., Laggsse, K., Luvivila, A.M., Ermans. 1977. 'Strategy of goitre and cretinism control in Central Africa'. *International Journal of Epidemiology*. 6: 43-5.

14

Iodine Deficiency Disorders and Iodized Salt for Livestock

M.G. Venkatesh Mannar and Chandrakant.S. Pandav

Abstract

The specific clinical disturbances caused by iodine deficiency in goats and sheep, beef and dairy cattle, horses, pigs, and poultry are discussed and an argument on economic lines for the prevention of IDD in farm animals is presented. Iodine supplementation through salt in the diet of farm animals is emphasized with a discussion of levels and toxicity. Providing salt to animals through the Universal Salt Iodization Programme is recommended.

Introduction

It is now well appreciated that Iodine Deficiency Disorders (IDD) are major public health problems in the world affecting large populations in over 110 countries with nearly 1.6 billion people at risk. What is not well known is the severe effect of iodine deficiency on animals. Farm animals are equally at risk to the spectrum of IDD at all stages of growth and development from conception to physical performance.

Unlike other nutrients such as iron, calcium or the vitamins, iodine does not occur naturally in specific foods; rather, it is present in the soil and imbibed through foods grown on that soil. Iodine is irregularly distributed over the earth's crust resulting in acute deficiencies in areas like mountainous regions and flooded riverines. The problem is aggravated by accelerated deforestation and soil erosion. This deficiency cannot be corrected. The food grown in iodine deficient regions can never provide enough iodine to the population and livestock living there. There are indeed rare examples of foods rich in iodine such as certain types of sea weed and sea fish; but these are not accessible to everyone. Thus, iodine deficiency results from geological rather than social and economic conditions. It cannot be eliminated by changing dietary habits or by eating specific kinds of foods. Rather the correction has to be achieved by supplying iodine from as an external source.

Effect of Iodine Deficiency in Farm Animals

Reproductive failure is often the outstanding manifestation of iodine deficiency and consequent impairement of thyroid activity in farm animals (Hetzel and Maberly, 1986). The birth of weak, dead, or hairless young in breeding stock has long been recognized in goitrous areas. Foetal development may be arrested at any stage, leading to either early death and resorption, abortion and stillbirth, or the live birth of weak young ones. Specific effects of iodine deficiency on different animals are subsequently discussed.

Goats and sheep

The influence of iodine supply on the reproduction of goats-success of conception, success of first insemination, abortion rate in gravid goats, length of gravidity, kids per gravid goat, and kids carried to term per gravid goat are in table 14.1. For example, the abortion rate in iodine deficient goats is 47 per cent as compared to 0 per cent in iodine sufficient goats. More kids are born to iodine sufficient goats and the probability of survival of kids born to iodine sufficient goats twice those that are born to iodine deficient goats. Similar results have been reported in sheep.

Table 14.1. Influence of iodine supply on the reproduction of goats

Description	Iodine content of dry matter in the feed (mg/kg)	
	0.04 (n=19)	0.40 (n=18)
Success of conception (%)	79	83
Success of first insemination (%)	27	73
Abortion rate of gravid goats (%)	47	0
Length of gravidity (days)	158	152
Kids per gravid goat	1.4	1.7
Kids carried to term per gravid goat	0.9	1.7

The symptoms of iodine deficiency in mature sheep are goitre, reduced yield of wool, reduced conception rate, or lambs born weak, dead, or without wool.

Beef and dairy cattle

Birth of calves with goitre is a sign of borderline or definite iodine deficiency although cows may appear normal and yet be afflicted. Iodine deficient calves may also be born blind, hairless, weak, or dead depending on the severity of the deficiency. Long-term deficiencies may result in decreased milk yields. In addition, irregular or suppressed estrus in dairy cattle causing infertility has been associated with goitre and shown to respond to iodine therapy. Work done by scientists in iodine deficient areas of Finland and Canada has demonstrated a marked improvement in the first service conception rate by feeding organic iodine preparations, beginning 8-10 days before the cows came into oestrus. Iodine deficiency is also known to cause male infertility (in bulls and stallions).

Horses

Pregnant mares may not show external signs of iodine deficiency but may produce a stillborn foal or one showing extreme weakness at birth. Foals born alive with a well developed goitre usually die or remain weak for the duration of their life.

Pigs

Piglets born to iodine deficient sows have thickened skin, puffy necks, hairlessness, and bloated appearance. Some piglets are born dead, while those born alive are weak and usually die within a few hours.

Poultry

Iodine deficiency in breeding hens results in reduced egg production, decreased hatchability, prolonged hatching time and thyroid enlargement in the embryos (March *et al.* 1984).

Economic benefits of Iodine Supplementation to Animals

Consumption of iodine by farm animals adds to their economic productivity. Adding trace minerals (iodine, copper, iron, cobalt, selenium, zinc and manganese) to animal diets benefits animal productivity. A study done at Purdue University with growing pigs showed that adding a trace mineral mixture to the diet improved gains by 15.4 per cent compared to the controls fed no trace minerals (Conrad *et al.* 1959). Another study showed that the cost of trace minerals was 4.7 US cents per pig but returned $1.64 US (a cost benefit ratio of 1:35). A study in Colombia indicated a return of 15.6 Pesos for each Peso invested in mineral supplements for animals compared to common salt supplements only.

The cost benefit ratio of iodine supplementation will be significant making the program a worthwhile investment, not only for the farm sector, but also for better human resource development.

Salt for animals

Both sodium and chlorine are necessary elements for animal life. In earlier times, the provision of salt in the rations of farm animals was haphazard, resulting in inefficient production (Bell, 1960). A continuously low salt intake affects the health of animals through loss of appetite and weight. Feed utilization decreases and more feed per unit of gain is required. In recent years, the role of common salt in animal feeding and nutrition has received increased attention. Salt is also a natural carrier for trace minerals since all animals need salt. Moreover, when cattle, horses, sheep and other animals graze on pasture with little, none, or varying amounts of concentrate feeding, farmers can always supply trace mineralized salt in a mineral block or box with free access, although this can lead to overconsumption of salt.

Salt as a carrier of nutrients

In the rural areas of many developing countries where iodine deficiency is most severe, people are largely dependent on subsistence foods. Their staple diet typically consists of one or two cereals and tubers or pulses. If a household has any livestock, dairy products may also be consumed. It is in this economic and dietary context that the choice of a vehicle for iodine must be considered.

Over the past 60 years, efforts to introduce iodine regularly in the daily diet have considered several foods as possible vehicles. Among these, salt has become the most commonly accepted, for a variety of reasons.

1) Salt is one of the few commodities that comes closest to being universally consumed by almost all sections of a community irrespective of economic status.

It is consumed at approximately the same level throughout the year in a given region by all normal adults. Thus, a micronutrient like iodine when introduced through salt can be administered to each individual at a uniform dosage throughout the year.

2) Compared to other food commodities whose production is widely dispersed, the production of salt is limited to fewer production centres. In many remote areas of the world, salt is one of the few commodities that comes from outside the local area, thereby lending itself to processing on an economical scale and under controlled conditions. By adding a fixed dosage of a micronutrient like iodine to salt at centralized locations, a majority of the population and livestock all over a region or country can ingest the nutrient with minimal effort of supplementation.

3) The mixing of an iodine compound with salt is a simple operation and produces no adverse chemical reactions. The equipment required is uncomplicated and easy to operate and maintain.

4) The addition of iodine to salt (usually as potassium or sodium iodide/iodate) does not impart any colour, taste or odour to the salt. In fact, iodized salt is indistinguishable from uniodized salt.

5) The cost of iodization is low: normally in the range 2 - 7 US cents per kilogram, which is less than 5 per cent of the retail price of salt in most countries.

Salt iodization is the long term and sustainable solution that will ensure that iodine reaches the entire population and is ingested on a regular basis. Once the iodization of salt is established as a permanent measure in a country, it eliminates iodine deficiency disorders and continues to provide each individual with a daily dose of iodine to prevent any recurrence. The introduction of iodine through salt has been successful in eliminating the problem of iodine deficiency in several countries for over 60 years. Since iodine is required in very minute quantities (150 - 200 micrograms), dosage in salt is extremely small. Salt consumption could be anywhere in the range of 5 - 20 grams a day within a given region or country. Normally, iodine concentration of salt is fixed in the range of 30 - 100 micrograms of iodine per gram of salt. This dosage takes into account anticipated iodine losses during transportation and storage.

Iodine supplementation for farm animals

In the developed countries, iodine supplementation had greatly reduced the incidence of IDD in both humans and animals. Reports from Montana, USA, indicate that before iodine feeding was practised there, goitre had caused an annual loss of thousands of pigs. Records from other areas show serious losses in the sheep and cattle industries prior to the discovery of the lack of iodine as the causative factor (McDowell, 1992).

Salt is the predominant vehicle for providing supplemental iodine for farm animals. Supplemental iodine may be incorporated into the salt, mineral mixture or concentrate feeds. For grazing animals in some inaccessible parts of the world, iodized salt blocks are air dropped. Salt iodized with either iodide and iodate are equally available to livestock. EIDD is an organic iodide used at relatively high dosages for the treatment of foot rot, lumpy jaw, and other conditions.

Iodine levels in salt for animals

In the developed countries, there are a range of trace mineralized salts that cater to each variety of livestock and poultry. In these trace mineralized salts, typically for livestock, levels of 70 -80 ppm iodine are used. Organized livestock rearing is a growing industry but is still limited. Farmers buy one type of salt from the market for both their livestock and their family.

Estimates of iodine levels in salt for different animals is presented in table 14.2. From this table, it can be seen that the iodine levels in salt varies over a wide range (i.e., 30 -400 ppm) and is species dependent. Minimum salt requirements for many categories of livestock can be ensured if the level of iodine in salt is approximately at least 20 ppm at the consumer level. If cattle and sheep continue to demonstrate iodine deficiency, then additional iodine supplementation may be required through cattle feeds or other sources. In all cases, however, there is no risk of toxicity to any class of animal, even with salt containing high levels of iodine of 200 ppm.

Table 14.2. Iodine requirements of animals

Class of Animal	Iodine Requirement in total Diet (ppm)	Maximum Dietary Tolerance Levels (ppm)	Salt Consumption (kg/year)	Iodine Required in salt (ppm)
Swine	0.14	400	4.1	28
Beef cattle	0.2 - 2.0	50	10	40 - 400
Dairy cattle	0.25 - 0.50	50	24.3	50 - 100
Horses	0.1	5	10.9	20
Sheep	0.1 - 0.8	50	4.1	20 - 160
Goats	0.15 - 0.80	50	4 - 8	30 - 240
Poultry	0.3 - 0.4	300	0.2	120 - 160

Source: *Cunha, T.J. Salt and Trace Minerals for Livestock, Poultry and Other Animals. The Salt Institute: Alexandria, Virginia, USA. 1987.*

Implications for the Universal Salt Iodization (USI) Programme

There is now a global thrust to eliminate iodine deficiency in humans through universal salt iodization (USI). A mid-decade goal of USI has been set for the end of 1995. Iodization programmes are already under way in over 60 countries. In order to have effective salt iodization programmes, all salt required for human and animal consumption must be iodized. If salt for human consumption is only iodized, then there exist a situation in the market where non-iodized salt, which is cheaper than iodized salt, is available for animal consumption. This leads people to purchase the cheaper, non-iodized salt. The availability of two types of salt also poses a major problem to law enforcing agencies as they cannot take legal action against those selling non-iodized salt since it is intended for animal consumption. In addition to issues of implementation, it is relevant to state here that farm animals are at equal risk of IDD as they reside in the same iodine deficient areas as humans.

Recommendations

1. Legislation covering the iodization of salt for human use should be extended to include all salt for livestock consumption. This will have the double benefit of ensuring that animals receive iodine supplementation and also that there is only one variety of salt in the market (iodized salt).

2. Advocacy and publicity campaigns should stress the need for iodine in animals and highlight the benefits of an iodine supplementation programme.

3. In many countries, it is very difficult to have separate grades of iodized salt for animals. Therefore, a separate monitoring of iodine intake of animals will be necessary to determine if certain categories require additional supplementation through feeds, etc.

References

Bell, T. Donald. 1960. 'Sodium Chloride in Animal Nutrition'. In: D.W. Kaufmann (ed.). *Sodium Chloride.* Reinhold Publishing Corporation: New York.

Conrad, J.H., M.P., Plumlee, W.M., Beeson. 1959. 'Trace mineral additions to corn soybean meal rations for growing-finishing pigs'. Purdue Agr. Exp. Sta. mimeo. AS-264.

Cunha, T.J. 1987. *Salt and Trace Minerals for Livestock, Poultry and Other Animals.* The Salt Institute: Alexandria, Virginia, USA.

Groppel, B., M., Anke, E., Scholz, B., Kohler.1986. 'The influence of different iodine supply on reproduction and the intrathyroidal iodine content of goats and sheep'.

Hetzel, B.S., G.F., Maberly. 1986. 'Iodine in Trace Elements'. In: *Human and Animal Nutrition.* Vol. 2. Academic Press: New York. pp. 139-208.

March, B.E., R.E., Austic, L.S., Jensen, D., Polin, J.L., Sell, P.E., Waibel P.W., Waldroup. 1984. *Nutrition Requirements of Poultry.* NAS-NRC: Washington, D.C.

McDowell, L.R. 1992. *Minerals in Animal and Human Nutrition.* Academic Press.

15

Iodised Salt Requirement for Livestock: A Case Study From Sikkim

R. Sankar

Sikkim is a small hilly state in the eastern Himalayas. Iodine deficiency disorders constitute a major public health problem in the state. Iodised salt programme has been in place in the state for the past one decade. The state receives iodized salt from Gujarat. The annual quota of salt is 5100 MTs. The population (1991) of the state is 4,06,457. Therefore the annual requirement should be 2440 MTs (6 kg/person/year). Perusal of records of department of Food and Civil Supplies reveal that the entire annual quota has been lifted and the salt sold in Sikkim for the past five years. Situational analysis of the iodised salt programme carried out in the state in the year 1994 had revealed that the salt requirement for livestock was high. The state has a livestock population of 3,32,177. The ratio of human to animal population is 1:0.8. A preliminary study was conducted to determine the salt requirement in different kinds of livestock. The study has shown that the salt requirement for livestock is high and that accounts for the sale of the entire quota of salt entering the state. There is a need to study the salt requirement in livestock in a systematic way. The annual quota of iodised salt for a state has to be worked out taking into consideration the livestock population.

•

Sikkim – Physical Aspects

Sikkim has an area of 7096 square kilometres. There are two main rivers, Rangit and Tista. The region is subjected to heavy rainfall, annual rainfall varying from 2000 to 5000 mm. The state is divided into four districts, namely, North, South, East and West.

Sikkim and its People

The population is distributed in 453 villages. Most of the population of the state live in small clusters; in 53.4 per cent of the villages, the population is less than 500 and there are only 6 towns in the state.

Animal Husbandry

Animal husbandry forms an integral part of the economy of Sikkim. A large percentage of the population is engaged in it in one form or the other. Livestock population of the state is given below:

Cattle	=	1,83,385
Buffaloes	=	3,088
Sheep & Goat	=	1,09,143
Yaks	=	5,354
Pigs	=	31,207

Prevalence of Iodine Deficiency Disorders

The entire state lies in the severely iodine deficient zone in the eastern Himalayas. Detailed epidemiological studies on the prevalence of iodine deficiency disorders were carried out in the state in 1990-1. These studies have shown that severe iodine deficiency exists almost everywhere in Sikkim. The overall goitre prevalence was found to be 54 per cent; prevalence of endemic cretinism 3.46 per cent; and mean urinary iodine concentration in a representative sample of the population was found to be 4 μg per decilitre.

Iodized Salt Programme

Iodized salt was introduced in the state in the late seventies. In 1985, necessary legislative measures were enforced to ban the sale of non-iodised salt in the state. PFA Act was enforced vide a notification issued in September 1985.

Annual quota of iodised salt for sikkim for the past five years has been:

1989 - 93	: 4250 MTs
1994 onwards	: 5100 MTs

The state has a well laid down 'Salt Movement Policy' and only iodized salt enters the state.

Iodized salt requirement of livestock—a preliminary study

Study Question

There is only one salt nominee in the state. The annual quota of salt has been lifted in full by the nominee for the past five years. Perusal of the salt nominee's records and the summary report of salt sold in the state, as provided by the Department of Food and Civil Supplies reveal that the entire annual quota of salt has been moved into Sikkim and sold there.

The salt requirement is calculated on the basis of 6 kg of salt per person per year. This is supposed to include the salt requirement of livestock. Therefore the annual requirement of salt for Sikkim for the population of 4,06,457 is 2440 MTs. There is no major industry in the state that uses salt. This raises the question:

WHERE DOES THE EXTRA SALT GO?

The possibilities considered were:

1. The annual quota of salt does not really enter the state in full or it is reexported.

2. The per capita salt requirement in this population may be higher than the Indian average figure of 10 grams per person per day.

3. The salt requirement for the livestock is more than what is presumed to be the case.

Perusal of the records of the department of Food and Civil Supplies reveals that the entire annual quota has been sold in Sikkim. The cost of iodized salt is more in Sikkim than in the neighbouring state of West Bengal. Therefore there is no incentive for the salt nominee to sell the salt outside Sikkim.

A nutrition survey was conducted in Sikkim in the years 1989-90. In this study the average salt consumption in Sikkim was found to range from 10 to 12 grams.

Fig. 15.1. Testing of iodized salt using a simple kit

A knowledge, attitude and practices (KAP) study was conducted in Sikkim in 1994 on the iodised salt practices in 1224 households. Of these households, 1030 (84.2%) households had livestock. In 897 households with livestock, the respondents said that they buy the same salt for human and animal consumption. In the remaining 13 per cent of the households with livestock, the respondents said they buy separate salt for animal and human consumption. 45 per cent of those interviewed said that they buy more than 30 kg salt at a time.

Preliminary inquiry revealed that in the households with livestock, nearly 80 per cent or more of the salt they were buying was going for feeding livestock. Therefore we proceeded to study the requirement of salt for livestock.

Salt Requirement for Livestock – a Preliminary Study

Fifteen households with livestocks were studied in the village of Namthang in the south district of Sikkim.

Name of Animals	Number of households with the animals
Cows	15
Pigs	5
Sheep/goat	6
Ox	3

A health worker visited these selected households and interviewed the persons who look after the livestock. They were questioned on the feeding practices of the livestock. Specific questions were asked on the quantity of salt that is fed to the animals. They were asked to show the quantity of salt that is fed to different animals. This was weighed with a balance.

The average quantity of salt fed to different livestock and the range as found in this study is given below:

Animal	Average salt consumption/day (gms)	Range (gms)
Cow	120	80-200
Pigs	40	30-60
Sheep/goat	30	20-50
Ox	400	350-550

There is no organized farm in Sikkim. Commercially available salt bricks and artificial solids are not used.

Using the above mentioned figures of average consumption of salt by different animals we have calculated the salt requirement for livestock in Sikkim and it is given below:

Animal	Salt/day	Salt for one year	No of animals in the state	Requirement/year
Cows	120 g	43.8 Kg	27,000	1183 MTs*
Pigs	40 g	14.6 Kg	31,207	455 MTs
Sheep Goat	30 g	10.9 Kg	1,09,143	1195 MTs
				2833 MTs

* The quantity of salt mentioned for cows is for milking cows. When a cow is pregnant and when for other reasons a cow is not lactating, the salt fed to such animals is said to be nearly one third

of what is fed to the milking cows. This calculation is based on the assumption that, 50 per cent of the cattle population in the state is cows and that 30 per cent of the cows are non-lactating at any given time.

The above calculation does not take into account the salt requirement for other animals such as yaks, bulls, buffaloes, etc.

Sikkim has a very large animal population. The ratio of human to livestock population in the country as a whole is likely to the lower than that seen in Sikkim (0.8).

The basis of calculation of 6 kg salt per person per year which is supposed to include the salt requirement of livestock is not clear. The results of our preliminary study show, (the poor methodology notwithstanding), that the salt requirement for livestock is high. The annual quota of salt allotted to a state has to take into account the livestock population of the state. For example, in Sikkim, the requirement is much higher than the annual quota because of the large livestock population. This can result in unscrupulous traders smuggling in uniodized salt into the state. There is urgent need to study the salt requirement for livestock in a systematic way. This will help in planning the iodized salt requirement for the country.

16

Iodization of salt for livestock in India

A.R. Rao

To ensure the consumption of iodized salt by the entire population of the country, the availability of the common uniodized salt should be prevented. Thus, iodized salt should also be available for the consumption of the total livestock in India.

Salt Requirements

Literature has been reviewed for data on the observed requirements of salt for various classes and ages of livestock, climates, terrain, feeds, fodders, etc.

Sodium requirements for growth ranging between approximately 0.1 per cent and 0.2 per cent of the ration have been reported from studies with rats, chicks, pigs, and calves.

Common salt: The age-old practice of including salt in the diets of both man and animals means that most of the sodium and chlorine is both ingested and excreted in this form. Since salt serves as a condiment as well as a nutrient, the intake tends to be highly variable and frequently in excess of needs. Its use as a condiment has physiological support in evidence that it stimulates salivary secretion and promotes the action of diastatic enzymes. When the intake is at a minimum, the body makes an adjustment whereby the output of sodium and chlorine in the urine nearly ceases. In contrast, large intakes involve a correspondingly large excretion, the water requirement being increased accordingly. The kidney is the regulating organ which, through its secretory activity, controls the concentration of electrolytes in the blood.

Dairy cows: In 1905 Babcock reported a long-time study of the role of salt in the dairy ration. He found that cows receiving no salt exhibited an abnormal appetite for it after two or three weeks, but that much longer time elapsed, even a year, before any ill effect on health was noted. Eventually there was a loss of appetite, an unthrifty condition, and a marked decline in weight and milk yield. These symptoms appeared first in the higher producers and the breakdown most frequently occurred at calving or shortly after the height of milk flow. The feeding of salt produced rapid recoveries in animals showing acute symptoms. The long period which elapsed before health was affected by salt deprivation illustrates the ability of the body to husband its supply of sodium and chlorine, reducing their excretion to a minimum, when the intakes are very small.

This early study was confirmed and extended, over fifty years later, by Smith and Aines in a much more comprehensive experiment with higher-producing cows. Sodium was found to be the primary deficiency. Its level in the blood plasma remained unchanged in cows receiving no salt, but the urine content dropped almost to zero within a month. The sodium, chloride, and potassium contents of milk remained unchanged. The supplemental salt (NaCl) required for lactating cows producing about 5000 kg of milk per year was estimated to be about 30 g per cow per day in addition to that in the feeds used, involving a total sodium requirement of about 21.3 g daily.

The human kidney may excrete as little as 1 g or as much as 40 g of sodium chloride per day, depending on the intake. Given normal kidneys and an appropriate water intake, large amounts can be excreted without harm. Excessive intakes, however, result in water retention in the body, causing edema. These results have been reported in chicks on rations containing little more than 3 percent of salt. Ruminants have a higher tolerance, but with a low water intake toxicity has been produced with a level of 2.2 per cent. It appears that the primary mode of action of salt poisoning is through a disturbance of water balance.

The salt requirement is greatly increased under conditions which cause heavy sweating because of the large loss in this secretion. Miners have been noted to lose 2½ kg of sweat per hour, containing 2 g of sodium chloride. If large amounts of water are drunk under these conditions, cramps results. The cramps disappear on drinking water containing salt. On a low-salt diet, however, the body gradually makes an adjustment whereby the concentration in the sweat, as well as in the urine, is gradually decreased. Equilibrium between intake and outgo is thus established at a much lower level than is possible initially.

Salt should be supplied regularly to cattle, or they will not thrive. Probably the best plan in the case of cows being fed the usual amount of a concentrate of 'grain' mixture is to mix 1.0 lb. of salt with each 100 lbs. of concentrates, and then in addition provide salt in a salt box or by the use of salt blocks, so the cows can always have access to it. Mixing some salt with the grain mixture tends to make it more palatable and also insures that the cows get salt each day.

Cows need about 0.75 ounce of salt daily per 1,000 lbs. live weight plus 0.3 ounce in addition for each 10 lbs. of milk they yield. If they are allowed to consume as much salt as they wish, they will take 1 to 4 ounces per head daily when fed ordinary dairy rations.

Beef cattle should have salt regularly, and the best method is to provide it where they can take what they can take when they wish, instead of salting them only once or twice a week.

For feeding where the salt is not exposed to the weather, loose salt, block salt, and rock salt are all satisfactory. Loose salt is not desirable for feeding in the open, because of the great loss through weathering. Cattle prefer the softer kinds of block salt to that which is very hard, and they usually eat somewhat more salt when fed loose salt then when supplied with block salt.

In 'Iowa' experiments over several years, cattle fattened in dry lot with shelter ate an average of about two-thirds pound of block slat per head per month. Where salt is fed in the open, about twice this amount must be supplied, to allow for wastage. In these tests the results were equally good from feeding block salt and loose salt.

Cattle on pasture consume much more salt than those fed in dry lot, and they eat more in spring and early summer, when the forage is abundant and succulent, then later in the season. In a Kansas test, yearlings and 2 year-old steers on pasture consumed about 2.8 lbs. of block salt per head in Juy, 1.8 lbs. in August, and

1.2 lbs. in September and October. Nearly equal amounts were lost by weathering of the salt blocks. In another 'Kansas' test, when a salt mixture was fed which was advertised as repelling flies from cattle, no such effect was noted.

Under range conditions a total of about 20 lbs. of salt is provided for each cow during the year. The salt allowance should be from 2.0 to 2.5 lbs. per head a month, when the feed is succulent or when the cattle are subsisting largely on browse. Later in the season, 1.0 to 1.5 lbs. per month are usually sufficient.

Sheep show a special fondness for salt and consume considerably more salt per 100 lbs. live weight than do cattle. Under any ordinary conditions, they should be furnished salt regularly. When sheep are accustomed to salt, it is the best plan to let them have access to it in suitable boxes or other containers, so they can take what they wish. Ewes will usually eat one-fourth to one-half ounce of salt a day, and fattening lambs one-fifth to one-fourth of an ounce.

In Iowa trials better results were secured from ewes fed one-fourth to one-half ounce of salt per head daily or allowed free access to salt, than when larger amounts were mixed with the feed or when no salt was supplied.

On the western ranges allowances of salt equivalent to one-fourth to one-half ounce per ewe daily (not counting the lambs) are considered reasonable. Often the sheep are salted at intervals of 3 to 7 days, instead of daily. In the alkali districts, sometimes range sheep are not salted, but merely allowed to eat alkali. This is safe, if the alkali contains 85 per cent salt.

Horses should be supplied regularly with salt. An allowance of 1.75 to 2.00 ounces per head daily is sufficient, and many horses eat less. A good plan is to supply salt where they can take what they wish, either loose salt from a suitable box, or block salt. Horses at hard work need more salt than others for a considerable amount is excreted in the sweat.

Although *swine* need less salt than do cattle, horses, or sheep, they should be supplied with it regularly, unless the feeds in the ration furnish enough. The best is to let them have access to salt in a suitable box or trough. When salt is thus fed, pigs will eat as much as 0.3 ounce per head daily.

Swine are apt to eat less block salt than flake salt. When block salt is used, it is best to include 20 percent of salt in the calcium-phosphorus mineral mixture, if such a mixture is also provided. When salt is mixed with the entire ration, not over one-half pound of salt should be added to each 100 lbs. of feed. If pigs have not had salt for sometime, they may at first eat too much if given free access to it.

When tankage, meat scraps, or fish meal is fed as the only or the chief protein supplement, there may be little or no benefit form supplying additional salt, because of the salt these feeds normally contain. On the other hand, it is necessary to provide salt for swine fed a ration made up wholly or almost entirely of feeds which are of plant origin.

The great benefit from adding salt to such rations is well shown by recent Indiana and Wisconsin trial. In 2 Indiana experiments pigs made an average daily gain of only 0.86 lb. and required 562 lbs. of feed per 100 lbs. gain when fed yellow corn, soybean oil meal, and a mineral mixture containing no salt. Other pigs supplied salt in addition gained 1.61 lbs. and needed only 262 lbs. of feed per 100 lbs. gain. The benefits from supplying salt was somewhat similar in a Wisconsin test.

Salt is needed by *poultry*, but the requirement is relatively small. It will be noted that in the nutrient allowances recommended by the committee of the National Research Council, the addition of 0.5 percent of salt to rations for chicks and hens is advised. The same percentage is recommended for *turkeys*.

If chickens are fed scratch grain and mash separately, the mash should contain about 1 percent of salt. When meat scraps, tankage, fish meal is used as the chief protein supplement, only 0.5 percent of salt need be added to the mash, because these feeds contain more salt than do the grains or the by-products of grains and other seeds.

Too great an amount of salt is injurious to poultry. However, chicks have grown normally on rations containing as much as 4 or 5 percent of salt, and even 8 per cent was not injurious in an Illinois experiment. On the other hand, 4 percent of salt was detrimental to turkey poults in a North Dakota trial, and poults died when fed rations containing 6 to 8 percent of salt.

The livestock population in India has been estimated from earlier recorded data, assuming the same rate of growth in their population that had been observed during the earlier years. These figure are given in table 16.1.

It has been estimated that the requirement of salt for the livestock in the country will be about 4.1 million tonnes per year by 1996.

Requirements of Iodization of Salt for Livestock

Information on the use of iodine supplements for animal husbandry is presented below:

Iodine Requirements: We do not have specific knowledge of the minimum iodine requirements of the various species, but we do know the levels that are at least high enough to prevent goiter. Goitre trouble in cattle, sheep, and hogs are associated primarily with reproduction. Although it appears that iodine needs must be greatest during reproduction, it is obviously required throughout life to keep the thyroid gland functioning normally. The appearance of goitre may result as an advanced stage of deficiency. On this basis the continued feeding of an additional supply might be considered desirable in areas of severe iodine deficiency. There is some evidence from certain areas that additional iodine improves growth in various species, but this evidence is not entirely convincing. It has also been reported that, with sows fed in dry lot in an area borderline with respect to the occurrence of goiter, additional iodine lessens the number of pigs dead at birth or which die in the first week, although there was little physical evidence of iodine deficiency. Recommendations for year-round feeding are usually put on an insurance basis or are suggested as the simplest way of making certain that the needs of breeding stock are taken care of in areas where goiter is very common. The amounts fed should be limited to recommended levels, for excesses can cause harm.

Table 16.1. Livestock populations in india and their requirements of salt (NaCl)

Species	Recorded No. in Million (Year)	Percent Annual Growth rate	Estimated No. in 1996 (million)	Salt required per day per head (g)	Salt required per year for 1996 Population (Mill.T.)
Cattle[a]	192.65 (1992)	-0.1 (1982-92)	191.88	30	2.10
Buffaloes[a]	78.55 (1992)	1.1 (1982-92)	82.00	30	0.90
Pigs[a]	10.50 (1992)	0.4 (1982-92)	10.67	5	0.02
Sheep[a]	44.41 (1992)	-1.2 (1982-92)	42.28	5	0.08
Goats[a]	117.00 (1992)	2.0 (1982-92)	126.36	5	0.23
Camels[b]	1.40 (1990)	2.04 (1951-90)	1.57	50	0.003
Horses[b]	0.90 (1990)	---	1.1	30	0.012
Asses and Mules[b]	1.60 (1990)	---	1.7	20	0.003
Chicken[a]	180.60 (1982)	9.5 (1982-92)	420.80		0.75[c]
			Total		4.098

Sources : [(a)] *RAPO (1993)*
[(b)] *FAO (1993)*
[(c)] *Assumptions : Two-thirds of the population are hens (layers consuming about 45 kg of feed containing 0.5% salt); the rest are broilers consuming 2.75 kg of feed with 0.5% salt for a lifetime of two months.*

Iodine supplements: Where the occurrence of goiter shows that attention to iodine nutrition is needed, the most practical method is the use of some special source such as iodized salt or sodium or potassium iodide. The addition of 0.0076 percent iodine to the salt represents the customary and effective level.

None of the common feedstuffs, with the exception of fish meal made from salt-water fish, can be relied upon to be rich in iodine. Dried kelp, a sea plant, is rich in the element, and so is cod-liver oil. While special advantages are sometimes claimed for organic sources of iodine, the preponderance of the evidence indicates that the cheaper inorganic iodides are equally satisfactory. When massive doses of iodine are given as a therapeutic agent, an organic source has the advantage of slower absorption and less risk of harm from overdosage, but in the amounts needed to prevent goiter, this is not a factor. Iodized salt loses its iodine rather readily under certain conditions because of the catalytic action of impurities present. Thus, after storage it may be an unreliable source of iodine unless stabilized. Some processes have been developed for accomplishing this stabilization. Iodized salt in mixed feeds is less subject to deterioration because the presence of proteins and unsaturated fats tend to stabilize the iodine. Both human and animal experiments have tested the usefulness of salt iodized with potassium iodate as a product of higher stability than where iodide is used. Davidson and coworkers have shown that salt blocks containing KIO_3 retained a much higher percentage of iodine when fed in the manger or subjected to outside exposure than did a product containing a commercial iodide complex.

Any required addition of iodine need not be supplied every day, because the thyroid has a considerable capacity to store the element. This is illustrated by the fact that goiter has been prevented in children in goitrous areas by feeding sodium iodide periodically for a month and repeating twice yearly.

Salt and iodine: While it is recognized as desirable to add salt to the rations of all farm animals, only limited information is available on the specific requirements for the growth of different species. Growth and balance studies to determine salt requirements have been carried out with pigs and chickens, as is illustrated by the report of Meyer and associates for swine. They concluded that the requirements are 53 to 64 mg of sodium daily per kilogram of body weight, amounts which would be provided by 0.3 percent of salt in the dry matter of the ration. The N.R.C. requirements call for 0.25 to 0.5 percent in the air-dry rations of both pigs and chickens. It is commonly recommended that lambs and calves have free access to salt.

The practice of allowing salt ad libitum to farm stock undoubtedly results in an intake in excess of requirements. There is no harm in this ordinarily, but certain species may be injured by overeating when salt is offered ad libitum after a long period of deprivation. Since salt increases the palatibility of rations, an intake in excess of the minimum physiological requirements is desirable from this standpoint. When it is fed free choice, the nature of the ration is a large factor governing the intake, and this may be in part a response to a variation in physiological need according to the mineral and other relations in the ration fed.

The NRC Report states that the iodine requirements for growing swine is somewhat less than 4.4 g per kilogram body weight. Supplementing the feed with 0.5 percent of iodized salt containing 0.007 percent of iodine will meet the requirement. For chicks the requirement is set at 0.35 mg per kilogram of feed. These levels for the chicks are higher than those found adequate for goiter prevention in other species. One reason for this high recommendation is to provide insurance against the goitrogenic effect of soybean meal, which is commonly used as a feed ingredient. While the growth needs for calves and lambs have not been established, it is clear that they do not exceed the amounts found adequate to prevent goiter troubles in reproduction in areas where these troubles otherwise occur. Harm can result from excessive intakes of iodine, as has been shown for lambs by Malan and associates.

In iodine-deficient areas, goitre is common in poultry, but production and health of the fowls is not generally affected appreciably. However, in such areas, it is wise to use iodized salt instead of common salt for poultry, or to include in the ration a feed such as fish meal, that serves as an iodine supplement.

Several experiments have been conducted in the USA and in Europe to find whether there was any advantage in adding iodine to ordinary poultry rations in other regions than those known to be deficient in iodine. Ewing concludes from a review of such experiments that in the experiments conducted in the USA there has been no evidence of a benefit to poultry from such an addition of iodine, and that in European experiments the data have been conflicting. Feeding a ration high in iodine definitely increases the iodine content of eggs.

Iodine requirements of different species under various conditions are given in table 16.2.

In view of the diversity of feeds and fodders, as also variations in them due to different seasons, stages of crop maturity and the numerous goitrogens, the requirement of iodization of salt for livestock has been enhanced over the years and currently the National Research Council of the National Academy of Sciences, USA recommends that the iodine content should be 190 ppm, rather than the earlier level of 76 ppm. It is often at a level of 200 ppm in other countries. Recent studies by my former colleagues at the Haryana Agricultural University, Hisar, have shown that about 15 mg of iodine supplied in about 30 g of common salt had marginal benefits in milk yields, even when the buffaloes were on good balanced rations. This concentration is conveniently possible when the level of iodization at the factory is about 200 ppm.

Implications of Feeding only Iodized Salt to Livestock

The biological implications of making iodized salt as the sole variety available for the consumption of livestock is very much like what is known for man. when iodine is supplemented, productivity is improved in proportion to the existing deficiency. However, some early reports suggested that the trouble of adding iodine was not worth while, when iodine deficiency was not present.

The economical implications of feeding only iodized salt to livestock is a different proposition, because of the enormous differences in the market prices

Table 16.2. Iodine requirements of various species

Species	Iodine Requirement	Reference
Human[1]	100-200	Curtis et at. (1937)
Human[1]	44-162	Scoffer (1933)
		Cole (1935)
Human, male[1]	100-200	Elmer (1938)
Human, male[1]	130-199	Wood (1970)
Human, pregnant female[1]	200	Wayne et al. (1964)
Human, children[1]	200	Wayne et al. (1964)
Rat[1]	0.7-0.9	Heinrich et al. (1964)
Rat[1]	20-40	Levine et al. (1933)
Poultry[1]	0.1-0.2	Wilgs et al. (1953)
Cattle, lactating[2]	0.8	Newton and Clawson (1974)
Pigs, growing[3]	0.14	Shuman and Townsend (1963)
Pheasants and quail[3]	0.30	Scott et al. (1960)
Dairy cattle, lactating[3*]	0.50	NRC, 1978
Dairy cattle, lactating[3*]	0.65	NRC, 1988
Dairy cattle, non-lactating[3*]	0.25	NRC, 1978
Dairy cattle, non-lactating[3*]	0.25	NRC, 1988

1 μg/day per person/animal
2 μg/kg Dry Matter of ingesta
3 PPM of Dry Matter of ingesta
* If the diet contains as much as 25% of a strongly goitrogenic feed on a dry matter basis, the amount of iodine provided should be increased two times or more.

Sources: *Curtis (1977), Scoffer (1977), Cole (1935), Elmer (1938).*

of common salt and the iodized variety. Recently, the two varieties were quoted at Rs 650 and Rs 3750 per tonne to the Department of Feed Technology at Hisar. The Salt Commissioner of the Govt. of India, however, states that the additional cost of iodization is approximately Rs 50 per tonne. If the difference in prices remains at the expected level, it is very much possible that livestock owners will be willing to pay this extra cost.

With the expected small differences in prices, the common farmer will be willing to use only iodized salt for feeding his animals, but the industry involved in the production of mineral mixtures, feed grain concentrates, etc. will try to use cheaper plain salts and additions of KI. However, the encouraging results of an increase by about 0.5 kg in the milk production in well-fed buffaloes of a good breed, shows that the extra cost of about 20 paise involved in feeding iodized salts, will easily be recouped. Of course, free access and extensive availability in various parts of the country is assumed.

Recommendations for Satisfactory Iodine Nutriture in Animals

Numerous measures have been adopted for ensuring iodine nutriture in man. Many of them can be adapted for animals:

1. Lugol's solution of iodine supplies 126 mg of iodine/ml. Similarly, tincture of iodine supplies about 146 mg of iodine/ml. So, a drop (0.05 ml) or two are enough/day/cow.

2. Tablets of potassium iodide to supply the same.

3. Potassium iodide in drinking water.

4. Potassium iodide in water administered by intramuscular injection periodically.

5. Iodine in unsaturated vegetable oils (poppy, groundnut, etc.) for administration orally at intervals of several months. However, there is no information about the changes in the oil or iodine by the micro-biota of the rumen in cattle and other ruminants.

6. Iodine in oil given intramuscularly at annual or longer intervals like in man. Injection of iodine in oil at intervals of one or more years is a satisfactory measure to tackle iodine deficiency disorders in livestock, through the regular field veterinary services (just as they do it for vaccinations against diseases).

7. Provision of iodized salts/solid salt blocks in the vicinity, for consumption (licking) by the animals at will. Potassium iodate is generally stable, but more expensive. Potassium iodide requires to be stabilised by the addition of 8% of calcium stearate.

8. Potassium iodide is usually added to mineral mixtures used as supplements in livestock rations. This is the usual method of supply of iodine in organised livestock enterprises in other countries as well as in India. However, there are few such in India and much of animal husbandry takes place as an auxiliary occupation of poor people.

9. Potassium iodide can also be added as a supplement in concentrate (grain) rations of livestock. This would be the most convenient way of supplying iodine either in the form of solution (as in 1 above) or as small tablets to be crushed into the concentrate (as in 2). However, many farmers do not make a regular habit of feeding even concentrates.

10. Iodine has been tried as a fertilizer for enriching foods, if feeds and fodders, though is not a necessary element for plants. However, different crops vary in their capacity to absorb and keep iodine in their tissues. While perennial rye grass may contain 146 g of I/100 g DM, other plants grown on the same soil may contain very low iodine levels of 10 g of I/100 g DM. However, modifications in agriculture do not appear to be promising in the prevention of iodine deficiency disorders.

11. Iodine in the form of Iodophors has been used often for the purification of drinking water. This is also an effective method of supplementing iodine. Levels less than 120 g of iodine/l of water do not affect taste. This method would be more expensive and difficult to reach remote areas.

12. Porous polymer vessels made of silicone elastomers have been used to keep iodine and release it slowly in community water supplies.

13. Long-acting intra-ruminal capsules, pots, etc. have been used to supply iodine as well as other trace minerals for periods of 2-3 years, since the rumen of livestock can hold solid objects for very long periods.

14. To supply iodine to the suckling young, the inside skin of cows' thighs have been painted occasionally/weekly with about 4 ml. of tincture of iodine or 2 ml. for pigs and sheep.

Thus, it is seen that numerous, diverse methods have been developed and implemented for the relief of iodine deficiency in various parts of the world.

In my opinion, methods 1 and 2 are adequate and inexpensive when supplemented with extension education of the farmers. Similarly, the 6th method mentioned above would be completely feasible by involving the veterinary departments. These would be inexpensive and would require much less of extra manpower or new organization.

For visibility, in small endemic areas, measure No. 13 would be effective, though at an unnecessary cost of about Rs. 250/- per cow, every two years.

Since the UNICEF is engaged at present in the promotion of iodized salt, it would be worthwhile to promote a more fortified variety for compulsory consumption as the sole source of sodium chloride for livestock also.

In Perspective

Iodine deficiency disorders of man have largely been tackled with supplements of oral iodized salt. It would no doubt be convenient to see that only such salt is available for animal feeding also, in the local markets, though it is very expensive. This is not the appropriate method for supplementing livestock with iodine. Due to increased costs, farmers may even reduce the feeding of salt.

Domestic animals require a higher level of iodine in their salt to counter the numerous and inestimable levels of goitrogens in their varying feeds over. The diverse terrain of a vast country like India. So, at least where care is taken to feed livestock, a little of potassium iodide may be directly added to the feeds, to minimize the costs of iodine supplementation.

Alternatively, periodic injections of iodine to large livestock populations are feasible through the Government veterinary departments, at low costs.

Such measures will certainly improve the health, fertility and productivity of livestock of even poor farmers.

References

A.I. Malan, P.J. du Toit, and J.W. Grownewald, 1935. Studies in mineral metabolism, XXXIII. Iodine in the nutrition of sheep, Onderstepoort J. Vet. Sci., 5: 189-200.

Babcock and Carlyle, Wis. Rpt. 1905

Becker, Neal, and Shealy, Fla. Bul. 231.

Bedi, S.P.S. 1995. - 'Iodine status of feeds and fodders and its role in livestock. Paper presented at the Technical Review Meeting on Iodine Deficiency Disorders and Livestock.' Pantnagar.

Cole, V.V. (1935). *J. Nutr.* 10:493.

Curtis, G.M., I.D., Puppel, V.V., Cole, N.L., Matthews. 1937 *J. Lab. Clin. Med. 22: 1014*, In Underwood, E.J. *Trace Elements in Human and Animal Nutrition*. 4th ed., Academic Press: New York. 1977.

Church D.C., and W.G. Pond 1983. - 'Basic Animal Nutrition and Feeding', 2nd ed., New York, Wiley.

Chapline and Talbot, U.S. Dept. Agr. Cir. 379; see also: Stanley, Ariz. Tech. Bul. 79; Crouch and Jones, Tex. Rpt. 1943.

Evvard et al. lowa Res. Bul. 94; Bayles, and Jones, Tex. Sta. mimco. rpt.; Madsen, and Phillips, Utah Bul. 292.

Evvard et al., lowa Res. Bul. 94.

Evvard, Breeder's Gazette, 76, 1919, p. 307; Evvard, Culbertson, et al., Iowa Sta., mimeo. rpts.; see also: Becker, Neal, and Shealy, Fla. Bul. 231; Jacob and Duncan, Tenn. Bul. 156.

Ewing, Poultry Nutrition, 3rd Ed., 1947, pp. 502-517.

Elmer, A.W. 1938. *Iodine Metabolism and Thyroid Function*. Oxford University Press : London and New York.
Foss, N.D. Bimo. Bul. 4, 1941, No.1.

FAO 1993. Production Year Book, vol. 46, Rome, Food and Agriculture Organization. Heinrich, H, G., Wenk, G., Bugdol. (1989) 'Effect of iodising mineral mix for cattle, sheep and pigs on iodine supply in man'. In: *6th International Trace Element Symposium*. (1989) 3: 756-62.

Hensel, Breeder's Gazette, 80, 1921, p. 181; Kan. Rpt. 1920-22.

Hudson, Mich. Quar. Bul., 8, 1926, No. 3; Templaton, Mis Rpt. 1928. Loeffel, Nebr. Rpt. 1924; Hale, Tex. Rpts. 1925, 1926; Sinclair, Sci. Agr., Canada, 20, 1939, p.p. 109-119.

Halpin, Holmes, and Hart, Wis. Bul. 425; Prentice, Jour. Ministry Agr. of North Ireland, 4, 1933, p. 72; Sjollema, Tierenahrung, 7, 1935, p. 184.

Hetzel B.S. 1989. *The Story of Iodine Deficiency*, Delhi, Oxford University Press.

Hetzel B.S. and C.S. Pandav. 1994 Eds. *S.O.S. for a Billion*, Delhi, Oxford University Press.

Jardine and Anderson, US Dept. Agr. Bul. 790; Sotola, Smith, and Ellington, Wash. Pop. Bul. 127.

J.H. Meyer et al. 1950. Sodium chloride and potassium requirements of growing pigs, J.Anim. Sci. 9: 300-06.

Kapoor V. and A.R. Rao 1995. -'Global occurrence of iodine deficiency in livestock'. Paper presented at the Technical Review Meeting on Iodine Deficiency Disorders and Livestock, Pantnagar.

Kapil U., K.S. Sohal, T.D. Sharma, D. Nayar, M. Goel, 'Consumption of salt by domestic animals' (in press).

Levine, H., Remington, R.E., Kolnitz, H.V. 1993. 'Studies on the relation of diet to goiter. 2. The iodine requirement of the rat'. *J.Nutr.* 6: 347-54.

Mannar M.G.V. and C.S. Pandav 1995. - 'Iodized salt for animal consumption'. Paper presented at the Technical Review Meeting on Iodine Deficiency Disorders and Livestock, Pantnagar.

Maynard, A. and K. Loosli 1979. - *Animal Nutrition*, 6th ed., New York, McGraw Hill.

Morrison, B. 1974. - *Feeds and Feeding*, 21st ed., ' Ithaca, Morrison Publishing Company', 1949. Newton, G.L., A.J., Clawson. *J. Anim. Sci. 39: 879.*

McCampbell et al., 1927. Fort Hays Branch, Kan. Sta., Cattlemen's Round-up.

NRC. 1971. 'Nutrient Requirements of Dairy Animals'. 4th ed., *National Academy of Sciences:Washington, D.C.*

NRC. 1978. 'Nutrient Requirements of Dairy Animals'. 5th ed., *National Academy of Sciences:Washington, D.C.*

NRC. 1980. 'Nutrient Requirements of Dairy Animals'. *National Academy of Sciences:Washington, D.C.*

NRC. 1988. 'Nutrient Requirements of Dairy Animals'. 6th ed. (revised) *National Academy of Sciences: Washington, D.C.*

Potter and Nelson, Western Live Stock Management, p. 146.

Pachauri S.P. 1995. -'Effect of iodine deficiency on livestock with special reference to the requirement of common salt and iodine'. Paper presented at the Technical Review Meeting on Iodine Deficiency Disorders and Livestock, Pantnagar.

Prakash R. 1995. - 'India leads in salt iodisation programme'. Paper presented at the Technical Review Meeting on Iodine Deficiency Disorders and Livestock, Pantnagar.

Prasad T. and K.S.N. Prasad. 1995. - 'Feed anitmetabolites which influence production of goitre in livestock'. Paper presented at the Technical Reveiw Meeting on Iodine Deficiency Disorders and Livestock, Pantnagar.

Quigley and Waite, Md. Bul. 343; Halpin, Holmes, and Hart, Wis. Bul. 425, and Poultry Sci., 15, 1936, p.99.

Radostits, O.M., D.C. Blood and C.C. Gay. 1989. - 'Veterinary Medicine', 8th ed., London, Bailliere Tindall.

RAPA. 1993. 'Selected Indicators of Food and Agriculture Development in Asia/Pacific Region', 1982/92, Regional Office for Asia and the Pacific (RAPA), F.A.O., Bangkok.

S.M. Babcock. 1905. The addition of salt to the ration of dairy cows, Wisconsin Agr. Expt. Sta. Ann. Rept. 22, pp. 129-156.

S.E. Smith and P.D. Aines. 1959. Salt requirements of dairy cows, Cornell Agr. Expt. Sta. Bull. 938.

Sachan R.S. and T.R. Rathore. 1995. - 'Current status of control measures for iodine deficiency in livestock'. Paper presented at the Technical Review Meeting on Iodine Deficiency Disorders and Livestock, Pantnagar.

Scoffer, L. 1933. Biochem.Z. 259: 11. In: Underwood, E.J. *Trace Elements in Human and Animal Nutrition*. 4th ed., Academic Press: New York. 1977.

Scott, M.L., Van Tien Loven, D.A., Holm, E.R., Reynolds, R.E.J. 1960 *Nutr. 70: 282.*

Shuman, A.C., Townsend, D.P.J. 1963. *Anim. Sci. 22:72,* In: Underwood, E.J. 'Trace Elements in Human and Animal Nutrition'. 4th ed Academic Press: New York. 1977.

Vestal, Ind. Sta:, mimeo. rpts. Grummer and Bohstedt, Wis. Bul. 472; see also: Kirk and Crown, Fla.Bul. 372; Evvard and Culbertson, lowa Leaflet 9 and mimeo. rpts.; Nebr. Rpt. 1924; Sinclair and Sackville, Sci. Agr., Canada, 6, 1926, pp.373-79.
W.M. Davidson, M.M. Finlayson, and C.J. Watson. 1951. Stability of various iodine compounds in salt blocks, Sci, Agr., 31:148-51.

Wayne, E.J., Koutras, D.A., Alexander, W.D. 1964. Clinical Aspects of Iodine Metabolism. Blackwell: Oxford.

Wilgs, H.S. Jr., Gassner, E.X., Patton, A.R., Harshfield, G.S. 1953. 'The iodine requirement of chickens'. Technical Bulletin. Colorado Agricultural Experiment Station. No. 49.

Wood, F.O. 1970. In: 'Iodine Nutrition in the United States'. *Natl. Acad. Sci. Natl. Res. Counc., Washington, D.C.* p.30.

PART VII

Recommendations

17

Technical Review Meeting on Iodine Deficiency Disorders in Livestock

Recommendations

1) Preparation of status report on Iodine Deficiency Disorders (IDD) in livestock in India:

This is to be achieved by collection and compilation of relevant data (both published and unpublished data) from various Indian Universities, individuals, and private organizations.

2) Development of database as follows:

2.1) Iodine deficiency disorders in livestock according to geographical regions

2.2) Existing and potential feeds and fodaers with iodine content

2.3) Institutions and individuals involved in iodine related work among:

2.3.1) livestock and poultry population

2.3.2) feeds and fodder

2.3.3) soil composition

3) Generation of programme related information

3.1) Biochemical indicators of IDD among livestock

To facilitate monitoring and impact evaluation of IDD control programme, there is a need to identify and standardize biochemical indicators of iodine deficiency among livestock. Since a reasonably well developed infrastructure to monitor IDD for human populations exists under the national IDD control programme, it will be efficient to develop indicators in livestock as well which require similar laboratory support.

The information so generated can be utilized to conduct cross-sectional studies among livestock of various regions and quantify the extent of iodine deficiency in the population.

3.2) Target species that can be used as markers of iodine deficiency among livestock.

Based on information generated as specified under para 3.1., livestock species can be identified that can be utilized as target animals for monitoring the level of iodine deficiency in the area. This information should be region specific.

4) Development of intervention strategies to combat IDD in livestock.

4.1) Preparation of a status report on global experience with intervention strategies to combat IDD in livestock populations.

4.2) Promotion of feeds and fodders that concentrate iodine.

There is a need to identify animal feeds that are locally available, palatable to the livestock and have high iodine content. Studies have shown that iodine content of the some fodders (e.g., barseem) vary with changing climatic conditions or the 'cutting number'. Similar studies for other commonly used fodders should be (frequency of fodder harvesting from same crops) carried out.

4.3) Iodine supplementation

4.3.1) Effect of iodized salt supplementation on iodine status of livestock

This strategy is similar to that adopted for human beings, However, the issues that need to be resolved are:

- Does livestock population receive salt in its daily diet as a routine in our country. If yes, how frequently and how much?
- What quantity of iodized salt (with 15 ppm iodine) will be required for the livestock to correct iodine deficiency and maintain appropriate body stores.

4.3.2) Iodine content and its stability in salt brick for licking by the livestock.

The National Dairy Development Board and some pharmaceutical companies are making iodine containing salt licking bricks. However, the availability of iodine from these to the animal is not known. Studies to determine the optimal iodine content of such bricks are necessary.

4.3.3) Development of mineral mixtures containing iodine and other micronutrients.

Micronutrients other than iodine are also deficient in the soil diets of livestock. Designing mineral mixtures appropriate for different livestock species and poultry keeping in view the Recommended Daily Allowances will be useful not only to correct iodine deficiency but also replace other micronutrients.

5) Other strategies to improve the iodine intake of young livestock

Iodophors are used in several developed countries to wash the udders of milch animals before milking. This increases the iodine consumption by the offspring (calves). An additional benefit is the increase in the iodine content of milk that, in turn is consumed by humans.

The availability and feasibility of its use in our settings and the economics of using iodophors should be studied.

6) To develop suitable communication strategies

Strategies for disseminating knowledge regarding iodine and other micronutrient deficiencies in the livestock population and its economic consequences to the farmer and the society.

The communication strategies have to be targeted towards:

-organized dairy and poultry farming

-individual farmers who manage the majority of cattle and poultry in this country

7) Develop models of cost benefit ratios

For iodine supplementation to livestock and poultry using different strategies.

8) Participation of relevant stakeholders

For the success of any programme to effectively correct iodine deficiency in the livestock, involvement of key stakeholders is essential.

Steps are required to encourage active participation of the following stakeholders:

- medical scientists
- veterinary scientists
- agricultural scientists
- government & public sector
- soil experts
- Industry
- Farmers co-operatives and NGOs

Index